POWER TO THE PEOPLE: THE ULTIMATE DIY SOLAR GUIDE FOR ENERGY INDEPENDENCE!

BY

Jon Springer

Evergreen Off-Grid AI Assist

evergreenoffgrid.com

This eBook is Presented by Evergreen Off-Grid AI Assist

Evergreen Off-Grid generated this eBook in part with GPT-3, OpenAI's large-scale language-generation model. Upon generating draft language, Evergreen Off-Grid reviewed, edited, and revised the language to our liking and we take ultimate responsibility for the content of this publication.

Table of Contents

[PREFACE] — GET READY TO HARNESS THE SUN!

In a single hour, the sun showers our planet with enough energy to power the entire world for an entire year. Imagine the possibilities if we could harness even a fraction of that immense power. Well, that's precisely what we're going to explore together in "Power to the People: The Ultimate DIY Solar Guide For Energy Independence."

So welcome, fellow solar enthusiasts! If you're here, you're likely intrigued by the idea of taking control of your energy needs and reducing your carbon footprint, all while saving money in the process. Solar power offers an accessible, clean, and sustainable energy solution, and this book will serve as your guide to creating your very own solar power system.

This isn't your ordinary technical manual. We've designed this book to be engaging and enjoyable, even for those who might not consider themselves experts in science or engineering. Our goal is to make solar power accessible to everyone, from the curious novice to the seasoned DIYer.

As you delve into this book, you'll discover everything you need to know about solar power systems, from the fundamental principles of harnessing the sun's energy to the intricacies of installation, maintenance, and troubleshooting. We'll be with you every step of the way with pro tips, and best practices to make your solar power journey a rewarding and enjoyable experience.

So, let's embark on this solar-powered adventure and unlock the incredible potential of our nearest star. Together, we'll demystify the world of solar power and empower you to take control of your energy future. It's time to harness the sun and start enjoying clean, sustainable energy.

[CHAPTER 1] — WELCOME TO THE SOLAR POWER UNIVERSE: A BEGINNER'S GUIDE

Greetings! You've embarked on an exciting journey, one that will illuminate the wonders of solar energy and reveal the secrets of harnessing the sun's mighty power for your own use. Whether you're an eco-conscious homeowner, a DIY enthusiast, or simply a curious individual seeking energy independence, you've come to the right place. Together, we'll unravel the mysteries of the solar power universe and learn how this incredible, clean energy source can impact our lives and the world around us.

In this chapter, we'll provide an overview of the solar power landscape, delving into the science behind converting sunlight into electricity, and exploring the components that make up a solar power system. We'll also discuss the remarkable benefits of solar power, touching on its environmental impact, financial savings, and potential for energy independence. Finally, we'll take a sneak peek at some practical solar power applications for homes and businesses and offer a few pointers on how to get started on your solar power journey.

This chapter is your launchpad into the world of solar energy. It's designed to be both informative and enjoyable, providing a solid foundation that you can build upon as you venture deeper into the subsequent chapters. So, let's embark on this exciting solar adventure and unlock the limitless potential of the sun!

THE SCIENCE BEHIND SOLAR POWER

Solar energy, a seemingly magical force, is powered by a series of fascinating scientific processes. To truly appreciate the remarkable nature of solar power, let's dive into the basics of how sunlight transforms into usable electricity.

The Sun: Our Cosmic Powerhouse

At the heart of our solar system, the sun is a massive ball of superheated gases, primarily hydrogen and helium. Through a process called nuclear fusion, the sun's core continuously converts hydrogen into helium, releasing an incredible amount of energy in the form of light and heat. This energy travels approximately 93 million miles to reach Earth, taking around 8 minutes and 20 seconds to make the journey. When sunlight reaches our planet, it becomes a valuable resource for generating electricity through solar power systems.

Photovoltaic Effect: The Key to Solar Power

The photovoltaic (PV) effect is the fundamental scientific principle that enables solar panels to convert sunlight into electricity. In simple terms, the PV effect occurs when photons (particles of light) collide with atoms in a solar panel's semiconducting material, usually silicon. This collision causes electrons to break free from their atoms, creating an electrical current.

Solar Cells: The Building Blocks of Solar Panels

Solar cells are the essential components responsible for converting sunlight into electricity. They are made of semiconducting materials, with silicon being the most common. When sunlight strikes a solar cell, the PV effect generates an electric current, which is then harvested to power electrical devices.

Each solar cell can only produce a small amount of electricity. Therefore, multiple solar cells are connected together to form a solar panel. By combining the output of many solar cells, a solar panel can generate a more substantial amount of electricity, suitable for a wide range of applications. Similarly, by combining many panels together you can form an array. Combining enough solar cells can power anything from a calculator to a city.

From Sunlight to Lightbulb

Solar panels convert sunlight into direct current (DC) electricity. However, most homes and businesses require alternating current (AC) electricity to power their electrical devices. This is where inverters come into play. An inverter is a crucial component of a solar power system that converts DC electricity generated by solar panels into AC electricity, making it compatible with the electrical grid and the devices we use every day.

In the upcoming sections, we're going to dive into the different parts of a solar power system and see how they work together, so we can learn how to design our own DIY system.

COMPONENTS OF A SOLAR POWER SYSTEM

Each component of a solar generation system plays a vital role in transforming sunlight into usable electricity. In this section, we'll briefly introduce the primary elements that make up a solar power system.

Solar Panels

Solar panels are the heart of any solar power system. They contain multiple solar cells made from semiconducting materials, which convert sunlight into direct current (DC) electricity using the photovoltaic effect. Solar panels come in various sizes, types, and efficiencies, which will be discussed in more detail in later chapters.

Inverters

Inverters play a crucial role in making the electricity generated by solar panels usable for powering everyday devices. They convert the DC electricity produced by solar panels into alternating current (AC) electricity, which is the standard form of electricity used by homes and businesses.

Charge Controllers

Charge controllers are essential components in solar power systems, particularly those with battery storage. They regulate the voltage and current flowing from the solar panels to the batteries, ensuring that the batteries are charged correctly and not overcharged, which can cause damage and reduce their lifespan.

Batteries

In off-grid and hybrid solar power systems, batteries are used to store the electricity generated by solar panels for later use. They allow the system to provide power during periods of low sunlight, such as at night or on cloudy days. There are various types of batteries available, each with their own advantages and disadvantages, which will be explored in later chapters.

Mounting Systems

Mounting systems provide the necessary support and structure to securely attach solar panels to a roof, ground, or other surfaces. They are designed to withstand various weather conditions and ensure that the panels are positioned at the optimal angle to maximize sunlight exposure.

Wiring and Electrical Components

A solar power system relies on wiring and electrical components, such as fuses, circuit breakers, and connectors, to safely and efficiently distribute the electricity generated by solar panels. These components help protect the system from electrical faults and ensure that the power is distributed correctly to the devices being powered.

In the upcoming chapters, we'll dive deeper into each of these components, learning about their functions, types, and how to select the right ones for your solar power system. With a solid understanding of these components, you'll be well-equipped to embark on your DIY solar project.

THE BENEFITS OF SOLAR POWER

Embracing solar power comes with a multitude of advantages that make it a wise choice for those looking to harness a renewable source of energy. In this section, we'll explore the key benefits that make solar power an attractive option for powering your home or business.

Environmental Impact

Solar energy is a clean and renewable source of power, with minimal greenhouse gas emissions and reduced dependence on fossil fuels. By choosing solar power, you contribute to a more sustainable future and help combat climate change.

Financial Savings

Over time, solar power systems can lead to significant savings on your energy bills. Although the initial investment can be substantial, the long-term financial benefits often outweigh the costs. Additionally, many regions offer incentives and rebates that can further reduce the overall expense.

Energy Independence

Utilizing solar power allows you to generate your electricity, reducing your reliance on the grid and increasing your energy independence. This can be particularly advantageous in case of power outages or fluctuations in energy prices.

Low Maintenance

Solar power systems are known for their low maintenance requirements. Once installed, they generally need minimal upkeep, with only occasional cleaning and periodic checks for any damage or wear. This makes solar power a reliable and hassle-free energy source.

Job Creation

The growing solar industry has led to the creation of numerous jobs in manufacturing, installation, and maintenance. By choosing solar power, you're not only investing in a cleaner energy source but also supporting a sector that contributes to economic growth and job opportunities.

Scalability and Flexibility

Solar power systems can be easily expanded or customized to meet your specific energy needs. Whether you're starting with a small setup or planning for future growth, solar power offers the flexibility to adapt to your unique circumstances.

Disaster Preparedness

Investing in solar power can provide an added layer of security and resilience during emergencies or natural disasters. In situations where the power grid is disrupted or damaged, having your solar power system, especially when combined with energy storage, can be lifesaving. It ensures that you maintain access to electricity for essential appliances and devices, increasing your self-sufficiency and reducing vulnerability during challenging times.

GETTING STARTED WITH SOLAR POWER

As you begin your solar power adventure, it's essential to familiarize yourself with some fundamental concepts and factors. In this section, we will provide a brief overview of some critical aspects to consider as you begin.

Assessing Your Energy Needs

What are you looking for? Are you powering a television or an air conditioner? Do you plan to power your entire home? Often, the first step in designing an efficient solar power system, is gaining a basic understanding of your energy consumption and needs.

If you plan to power your entire home, you'll need to know your average monthly usage and consider any potential changes in your energy needs in the future.

Identifying your energy needs will serve as a foundation for designing your solar power system later on.

Evaluating Your Location for Solar Potential

The location of your solar power system plays a significant role in its overall effectiveness. Factors such as sunlight exposure, roof orientation and angle, and potential shading can impact the performance of your solar panels. In the upcoming chapters, we will delve into the details of evaluating your location for optimal solar potential.

Understanding Solar Incentives and Policies

Solar incentives and policies are essential components of the solar power landscape. They can help make solar power systems more affordable and accessible by offering financial benefits like tax credits, rebates, and grants. As we progress through the book, we will discuss various incentives and policies available and how you can take advantage of them.

Understanding Payback Period

The payback period is an important financial metric to consider when investing in a solar power system. It represents the amount of time it takes for the savings generated by the system to equal the initial investment cost. After the payback period, the system starts generating net savings or profit.

Here's a closer look at the factors that influence the payback period and how to calculate it for your solar power system.

Factors Affecting the Payback Period

Several factors can impact the payback period of a solar power system, including:

• System cost: The initial cost of purchasing and installing the solar power system is a critical factor in determining the payback period. Lower system costs will result in shorter payback periods.

• Energy savings: The amount of money saved on electricity bills by generating solar power directly affects the payback period. Greater energy savings lead to shorter payback periods.

• Incentives and rebates: Financial incentives, such as tax credits, rebates, and grants, can significantly reduce the initial cost of a solar power system, shortening the payback period.

• Maintenance costs: Although solar power systems are generally low-maintenance, any costs associated with maintaining the system should be factored into the payback period calculation.

Calculating the Payback Period

To calculate the payback period for a solar power system, follow these steps:

Step 1: Determine the net system cost. Subtract any incentives, rebates, or grants from the initial system cost to obtain the net system cost.

Step 2: Calculate the annual energy savings. Estimate the amount of electricity your solar power system will generate in a year and multiply it by the cost of electricity from the grid. This will give you the annual energy savings.

Step 3: Factor in maintenance costs. Estimate any annual maintenance costs for your solar power system and subtract them from the annual energy savings calculated in step 2.

Step 4: Calculate the payback period. Divide the net system cost (step 1) by the adjusted annual energy savings (step 3) to obtain the payback period in years.

A shorter payback period indicates a faster return on investment and is generally more attractive to potential solar power system owners. By understanding and calculating the payback period, you can make more informed decisions about the financial feasibility of investing in solar power.

EMBRACING THE POWER OF THE SUN: YOUR JOURNEY AWAITS!

As we wrap up this introductory chapter, it's essential to recognize that we've only just scratched the surface of the incredible world of solar power. With each turn of the page, you'll continue to uncover the fascinating intricacies of harnessing the sun's energy, empowering you to take control of your power needs and contribute to a more sustainable future.

Together, we will explore the ins and outs of solar power systems, learning how to evaluate your energy needs, choose the right components, and install your very own solar power setup. You'll also gain valuable insights into maximizing your solar savings, maintaining your system, and understanding the policies and incentives that can make solar power more accessible than ever before.

As you embark on this enlightening journey, remember that you are not alone. We are here to guide you every step of the way, providing you with the knowledge and tools you need to succeed in the realm of solar power. So, buckle up and get ready to immerse yourself in the thrilling adventure that lies ahead! The sun's energy is vast and powerful, and together, we will learn how to harness it for a brighter, more sustainable world.

[CHAPTER 2] — SUNBEAMS AND ELECTRONS: DEMYSTIFYING THE BASICS OF SOLAR POWER

In this chapter, we'll unravel the science that transforms sunlight into clean, renewable energy. We'll uncover the marvels of solar power and its potential to revolutionize the way we power our lives.

We'll uncover the basics of how sunlight becomes electricity and the key processes involved in making that happen. From the awe-inspiring nuclear fusion brewing within the sun to the physics behind photovoltaic cells, we'll take you on a captivating exploration of the principles that make solar power possible.

So, grab a cup of your favorite beverage, find a comfy spot, and join us as we explore the science behind harnessing the sun's energy.

SOLAR RADIATION: A CONSTANT FLOW OF ENERGY

The Electromagnetic Spectrum

Before we dive into the specifics of solar energy composition, it's essential to have a basic understanding of the electromagnetic spectrum. The electromagnetic spectrum is a range of all types of electromagnetic radiation, which includes everything from radio waves to gamma rays.

Electromagnetic radiation is nothing but a form of energy, and visible light represents just a tiny portion of this immense spectrum. The spectrum is generally divided into sections based on wavelength, which is the distance between successive peaks or troughs of a wave.

The sections, ordered from longest to shortest wavelengths, include radio waves, microwaves, infrared light, visible light, ultraviolet light, X-rays, and gamma rays.

Solar Energy Composition: Ultraviolet, Visible, and Infrared Light

The sun emits a broad range of electromagnetic radiation, but the majority of solar energy reaching Earth consists of three main types: ultraviolet (UV) light, visible light, and infrared (IR) light.

Ultraviolet Light: Ultraviolet light has shorter wavelengths than visible light and makes up about 10% of the sun's total energy output. Though UV light represents only a small fraction of the sun's energy, it is responsible for some crucial processes on Earth, such as the production of vitamin D in our bodies and the formation of the ozone layer. However, excessive exposure to

UV light can also have harmful effects, such as sunburns and skin cancer.

Visible Light: The visible light spectrum is the range of wavelengths that our eyes can detect, and it accounts for about 43% of the sun's total energy output. Visible light is essential for life on Earth, as it provides the energy needed for photosynthesis in plants and helps maintain the planet's temperature.

Infrared Light: Infrared light has longer wavelengths than visible light and accounts for approximately 47% of the sun's energy output. Infrared radiation is responsible for the warmth we feel from the sun, and it plays a vital role in the Earth's climate system. While we can't see infrared light, we can feel it as heat, and special instruments can detect its presence.

To sum it up, the sun's energy primarily consists of ultraviolet, visible, and infrared light, each having distinctive attributes and influences on Earth. While delving into the realm of solar power, it's crucial to grasp how these solar radiation elements interact with the technology we employ to capture the sun's energy and transform it into renewable electricity.

HOW SOLAR ENERGY REACHES EARTH'S SURFACE

The Earth's Atmosphere: A Solar Radiation Filter

One of the most critical factors influencing the amount of solar energy reaching the Earth's surface is the atmosphere. As sunlight travels through the atmosphere, it interacts with various gases, aerosols, and clouds. These interactions can cause solar radiation to be absorbed, scattered, or reflected, changing the energy levels that ultimately reach the surface.

Direct and Diffuse Sunlight: Two Sides of the Solar Coin

When discussing solar radiation, it's essential to understand the distinction between direct and diffuse sunlight.

Direct sunlight is the solar radiation that travels straight from the sun to the Earth's surface without any interference from the atmosphere. This light is more intense and concentrated, making it the primary source of energy for solar panels.

On the other hand, diffuse sunlight is the solar radiation that has been scattered, reflected, or absorbed and then re-emitted by atmospheric particles, such as dust, aerosols, and water vapor. This light is less intense and spread more evenly across the sky, contributing to a smaller portion of the energy collected by solar panels.

Navigating the Influencers: Factors Affecting Solar Energy Availability on Earth's Surface

Several factors can influence the amount of solar energy reaching the Earth's surface. Some of these factors include:

1. Latitude: Solar radiation levels vary depending on the Earth's latitude. As you move closer to the equator, solar energy levels increase due to the more direct angle of sunlight.
2. Time of day and year: The position of the sun in the sky changes throughout the day and year, affecting the angle at which sunlight strikes the Earth's surface. Solar radiation is typically most intense around solar noon when the sun is at its highest point in the sky.
3. Weather conditions: Cloud cover, precipitation, and atmospheric pollution can all impact the amount of solar energy that reaches the Earth's surface. In general, cloudy or polluted conditions result in lower solar radiation levels.
4. Surface elevation: Higher elevations often receive more solar radiation due to the thinner atmosphere and reduced scattering and absorption of sunlight.

By understanding these factors and how they influence solar energy levels, you can better assess your location's solar potential and optimize your solar power system accordingly.

PHOTOVOLTAIC EFFECT: THE HEART OF SOLAR POWER

A Glimpse into the Past: The Discovery of the Photovoltaic Effect

Our journey to understanding the heart of solar power takes us back to the early days of scientific discovery. It was the 19th century, a time of great innovation and curiosity, when the foundations of modern solar power were first established. The photovoltaic effect, which would later revolutionize the way we harness energy from the sun, was discovered by a young French physicist named Edmond Becquerel.

Born into a family of esteemed scientists, Edmond Becquerel was destined for greatness. His father, Antoine César Becquerel, was a well-known scientist who made significant contributions to the field of electrochemistry. Following in his father's footsteps, young Edmond started conducting experiments at the tender age of 19.

In 1839, while experimenting with an electrode submerged in a conductive solution, Becquerel stumbled upon a phenomenon that would change the course of history. He observed that when the electrode was exposed to sunlight, it generated a voltage. This groundbreaking observation marked the birth of the photovoltaic effect and laid the foundation for future developments in solar power technology.

Becquerel's discovery, however, did not immediately spark a solar revolution. It would take more than a century for scientists to fully understand the implications of his findings and develop efficient solar cells. Yet, Becquerel's work provided the vital spark that ignited the flame of innovation, inspiring future generations of researchers to explore the potential of solar energy.

Today, as we harness the sun's power to light our homes, charge our devices, and fuel our vehicles, we stand on the shoulders of giants like Edmond Becquerel. His pioneering discovery of the photovoltaic effect opened the door to a world of clean, renewable energy, paving the way for a more sustainable future for our planet.

Cracking the Code: The Physics Behind the Photovoltaic Effect

The photovoltaic effect is a physical phenomenon that occurs when light strikes a material, causing it to absorb photons and release electrons. These free electrons can then be captured and directed into an electrical current. The key to this process lies in the ability of certain materials to absorb and convert sunlight into usable energy.

Semiconductors: The Unsung Heroes of Solar Cells

Enter the world of semiconductors, the real heroes of solar power. Semiconductors are unique materials that possess properties of both conductors and insulators, allowing them to control the flow of electrical current. The most commonly used semiconductor in solar cells is silicon, which has a crystalline structure that forms an ideal environment for the photovoltaic effect to take place.

When sunlight hits a solar cell, the energy of the absorbed photons excites the electrons in the silicon atoms, causing them to break free from their atomic bonds. This process creates a surplus of free electrons on one side of the solar cell and a shortage on the other side. By connecting the two sides with an external circuit, the electrons flow, generating an electrical current that can be used to power our homes, gadgets, and more.

In a nutshell, the photovoltaic effect is the beating heart of solar power, transforming sunlight into clean, renewable energy. Grasping the underlying principles of this extraordinary process allows us to recognize and harness solar power's immense potential.

SOLAR CELL EFFICIENCY: GETTING THE MOST OUT OF SUNLIGHT

Factors Affecting Solar Cell Efficiency

Solar cell efficiency refers to the percentage of sunlight that is converted into usable electricity by a solar cell. Several factors can impact a solar cell's efficiency, including:

1. Temperature: Higher temperatures can reduce the efficiency of a solar cell, as the increased heat can cause greater energy losses.
2. Angle of incidence: The angle at which sunlight hits the solar cell can impact its efficiency, with a direct angle being more efficient than an indirect angle.
3. Material quality: The quality of the materials used in the solar cell, including the semiconductor and other components, can affect its efficiency.

TYPES OF SOLAR PANEL TECHNOLOGIES

Monocrystalline Silicon Panels

Monocrystalline silicon panels are made from a single crystal structure and have a uniform dark appearance. These panels offer the highest efficiency, typically between 18% and 22%, and are generally more expensive than other types of solar panels.

They are also more space-efficient, making them an ideal choice for smaller installations or areas with limited space.

Polycrystalline Silicon Panels

Polycrystalline silicon panels are made from multiple crystal structures and have a blueish hue. They have slightly lower efficiencies than monocrystalline panels, usually coming in between 15% and 18%, but are typically less expensive. These panels are a popular choice for residential installations due to their balance of efficiency and cost.

Thin-Film Solar Panels

Thin-film solar panels are made by depositing layers of photovoltaic material onto a substrate. These panels are lightweight and flexible, making them suitable for a variety of applications, such as building-integrated photovoltaics. However, their lower efficiency, typically between 10% and 12%, means they require more surface area to produce the same amount of power as traditional silicon panels.

Emerging Solar Panel Technologies: Innovations and Future Potential

The solar industry is constantly evolving, with researchers and engineers developing new technologies to push the boundaries of efficiency, cost-effectiveness, and versatility. Here are some emerging solar panel technologies that show significant potential for transforming the way we harness the sun's energy:

Perovskite Solar Cells: Revolutionizing Efficiency and Cost

Perovskite solar cells are made from a unique class of materials that offer high efficiency at a lower cost compared to traditional silicon cells. These cells have rapidly progressed in recent years, with laboratory efficiencies now rivaling those of silicon-based technologies. They can also be manufactured using more cost-effective and scalable methods, making them an attractive option for future large-scale solar power deployments.

Organic Photovoltaics: Flexible, Lightweight, and Customizable

Organic photovoltaics (OPVs) use organic materials, such as polymers or small molecules, to capture sunlight and convert it into electricity. One of the most appealing aspects of OPVs is their potential for flexibility, lightweight design, and customization. This technology could enable solar panels with adaptable colors, transparency, and form factors, opening up new possibilities for integrating solar energy into various applications, such as building facades, windows, and even clothing.

Building-integrated photovoltaics (BIPV):

BIPV technology involves incorporating solar panels directly into building materials, such as glass, roofing, or facades. This approach not only saves space but also adds aesthetic appeal and generates clean energy seamlessly. With advancements in materials and design, BIPV is poised to become an integral part of the construction industry in the coming years.

Quantum Dot Solar Cells: Tapping into the Full Solar Spectrum

Quantum dot solar cells employ tiny particles called quantum dots to capture sunlight and convert it into electricity. These cells have the potential to achieve higher efficiencies by capturing a broader range of the solar spectrum compared to traditional solar cells. Quantum dots can be tuned to absorb specific wavelengths of light, allowing for the design of solar cells that can utilize a larger portion of the solar spectrum. This technology holds promise for significantly improving the performance of solar panels and expanding the range of applications for solar power.

SOLAR POWER SYSTEM PERFORMANCE METRICS

Solar Panel Efficiency

Solar panel efficiency is a measure of how effectively a solar panel can convert sunlight into usable electricity. It is expressed as a percentage, representing the proportion of sunlight that is converted into electricity by the panel. Higher efficiency panels can generate more electricity from the same amount of sunlight, making them more space-efficient and potentially cost-effective in the long run.

Temperature Coefficient

The temperature coefficient is a measure of how a solar panel's performance is affected by changes in temperature. It is typically expressed as a percentage per degree Celsius (e.g., -0.4%/°C). A lower (less negative) temperature coefficient means the panel's performance is less affected by temperature increases. This is an important consideration for solar power systems in areas with high temperatures or significant temperature fluctuations, as higher temperatures can reduce a panel's efficiency and output.

Performance Ratio

The performance ratio is a metric used to evaluate the overall performance of a solar power system, taking into account factors such as energy losses, temperature effects, and shading. It is calculated by comparing the actual energy output of a solar system to the expected energy output under ideal conditions. A higher performance ratio indicates that the solar power system is operating closer to its maximum potential, while a lower ratio indicates that there may be inefficiencies or issues affecting the system's performance.

Energy Payback Time

Energy payback time (EPBT) is a measure of the time it takes for a solar power system to generate enough energy to offset the energy used in its production, installation, and maintenance. This metric helps to evaluate the environmental impact and sustainability of solar power systems. A shorter energy payback time indicates that the system will produce more energy over its lifetime compared to the energy used in its production and operation. Factors that influence EPBT include the efficiency of the solar panels, the location of the installation, and the type of solar power system used.

CONCLUSION: UNDERSTANDING THE FOUNDATIONS FOR A BRIGHTER FUTURE

As we conclude our exploration of the basics of solar power, it's crucial to understand the importance of grasping these fundamental concepts. Knowledge of solar power technology empowers enthusiasts and potential system owners, helping them make informed decisions about adopting this clean and sustainable energy source.

The future of solar power technology holds immense potential, and the innovations we discussed are just the beginning. As the industry continues to grow and evolve, we can expect to see even more groundbreaking advancements that will further harness the power of the sun, reduce our reliance on fossil fuels, and contribute to a cleaner, more sustainable planet.

As we close the chapter on the science of solar, we hope you have gained valuable insights into the intricate workings of this powerful, renewable energy source. In the upcoming chapter, we'll shift our focus to assessing your energy needs. Getting this part right is crucial to designing an efficient solar power system tailored to your unique requirements. So, get ready to assess your energy needs and power up your life .

[CHAPTER 3] — POWER UP YOUR LIFE: ASSESSING YOUR ENERGY NEEDS

LAYING THE FOUNDATION FOR A CUSTOMIZED SOLAR POWER SYSTEM

Embarking on your solar power journey is an exciting endeavor, but it's important to start with a solid foundation. This foundation is rooted in understanding your energy needs and how they relate to the size of your solar power system. By tailoring your solar power system to meet your specific energy needs, you'll be able to maximize its performance, reap the benefits of energy independence, and optimize your financial returns.

The Link Between Energy Needs and Solar Power System Size

One crucial aspect of designing a solar power system is ensuring it is the right size to meet your energy needs. A system that is too small may not produce enough power to cover your electricity requirements, leaving you reliant on the grid and possibly incurring additional costs. On the other hand, a system that is too large can be unnecessarily expensive to install and may lead to excess energy production that goes unused.

How Tailoring a Solar Power System to Your Needs Maximizes Performance

By customizing your solar power system based on your energy needs, you ensure optimal performance and cost-effectiveness. This allows you to reap the full benefits of solar power, from reducing your carbon footprint to minimizing your reliance on the grid, all while getting the most out of your investment.

READING BETWEEN THE LINES: DECODING YOUR ELECTRICITY BILLS

Key Components of Your Utility Bill

To accurately assess your energy needs, you'll first need to understand your current electricity consumption. This information can be found on your utility bill, which typically includes:

- Total energy usage (in kilowatt-hours or kWh)
- Billing period duration
- Cost per kWh
- Any additional charges or fees

Calculating Your Average Monthly Energy Usage

To determine your average monthly energy usage, gather your utility bills from the past year. Add up the total energy usage (in kWh) and divide this number by 12 (the number of months in a year). This will give you an approximation of your average monthly energy consumption, which is an essential piece of information when designing your solar power system.

Recognizing Patterns and Adapting to Seasonal Changes

As you examine your utility bills, you may notice patterns in your energy consumption. These patterns could be due to seasonal changes, such as increased air conditioning use in the summer or heating in the winter. By understanding these patterns, you can better anticipate your energy needs and ensure your solar power system is designed to accommodate fluctuations in your energy consumption throughout the year.

PLANNING FOR TOMORROW: ESTIMATING YOUR FUTURE ENERGY NEEDS

Designing a solar power system that meets your current energy needs is essential, but it's also important to consider how your energy consumption may change in the future. By factoring in potential lifestyle changes, home improvements, and evolving technology, you can create a solar power system that will continue to meet your needs for years to come.

Factoring in Lifestyle Changes and Home Improvements

As you assess your future energy needs, consider any lifestyle changes that may impact your energy consumption. Are you planning to expand your family, or do you anticipate your children moving out in the near future? Are you considering working from home, which could increase your daytime energy usage? Additionally, think about any planned home improvements, such as additions or renovations, that may affect your energy needs.

Accounting for New Appliances and Electric Vehicles

Technology is constantly evolving, and new, energy-consuming devices and appliances are continually entering the market. As you plan your solar power system, consider how your energy needs might change with the addition of new appliances or devices. For example, it's looking like we will all be purchasing an electric vehicle (EV) in the not too distant future, so you'll need to account for the additional energy required to charge it. Be sure to communicate these potential changes to your solar installer or into your design plans to accommodate your future energy requirements.

The Role of Energy Efficiency in Reducing Future Energy Consumption

Embracing energy efficiency measures can help you minimize your future energy needs and make it easier to size your solar power system appropriately. Upgrading to energy-efficient appliances, installing a programmable thermostat, and making other energy-saving home improvements can help reduce your overall energy consumption, making your solar power system even more effective in meeting your needs.

By taking the time to consider your future energy needs and making adjustments to your solar power system design accordingly, you can ensure that your investment in solar energy continues to pay off for years to come.

HOME ENERGY AUDIT: IDENTIFYING OPPORTUNITIES FOR ENERGY SAVINGS

Conducting a DIY Home Energy Audit

A home energy audit is an essential step in identifying areas where you can save energy and reduce costs. Performing a do-it-yourself (DIY) audit can help you understand your energy consumption patterns and find opportunities for improvement.

1. Inspect your home's insulation: Check the attic, walls, and floors for adequate insulation. Insufficient insulation can

lead to heat loss in winter and heat gain in summer, increasing your energy consumption.

2. Examine windows and doors: Look for drafts, gaps, and cracks around windows and doors that may allow air to escape or enter your home. Sealing these gaps can significantly improve your home's energy efficiency.
3. Evaluate your heating and cooling system: Check the age and condition of your HVAC system. Older systems are typically less efficient and may need to be upgraded to save energy.
4. Assess your lighting: Replace incandescent bulbs with energy-efficient LED or CFL bulbs to reduce electricity usage.

Recognizing Common Sources of Energy Waste

By understanding the common sources of energy waste, you can take steps to reduce your energy consumption and save money. Some of the most prevalent sources of energy waste include:

1. Inefficient appliances: Old or poorly maintained appliances can consume more energy than necessary. Upgrading to Energy Star-rated appliances can lead to significant savings.
2. Vampire power: Electronic devices and chargers that are plugged in but not in use still consume energy. Unplug

these devices or use power strips with an on/off switch to reduce this energy waste.

3. Poorly sealed ductwork: Leaks in your home's ductwork can result in wasted energy as heated or cooled air escapes before reaching its destination.
4. Excessive water heating: Heating water accounts for a large portion of your home's energy usage. Reducing the thermostat setting on your water heater and insulating the tank can help save energy.

Implementing Energy Efficiency Measures to Lower Energy Needs

Once you've identified areas where you can save energy, implement the following measures to improve your home's efficiency:

1. Seal gaps and cracks: Apply weatherstripping or caulk to seal any gaps around windows and doors to prevent drafts.
2. Install a programmable thermostat: Automatically adjusting your home's temperature based on your schedule can lead to energy savings.
3. Improve insulation: Add insulation to your attic, walls, and floors to maintain a comfortable indoor temperature with less energy consumption.

4. Use energy-efficient appliances: Replace old appliances with Energy Star-rated models that consume less energy.
5. Maintain your HVAC system: Regularly clean and replace filters, and schedule annual maintenance to keep your heating and cooling system running efficiently.

By conducting a home energy audit and implementing these energy-saving measures, you'll be well on your way to reducing your energy needs and making your home more efficient.

SIZING YOUR SOLAR POWER SYSTEM: MATCHING PRODUCTION TO CONSUMPTION

The Importance of Accurate Energy Needs Assessment

Accurately assessing your energy needs is crucial for designing a solar power system that meets your consumption requirements. By thoroughly evaluating your energy usage patterns, you can determine the right system size to generate the necessary amount of electricity without oversizing or undersizing your solar array. Proper system sizing can enhance your system's efficiency, maximize your return on investment, and minimize the chances of overloading or underutilizing your solar panels.

Balancing System Size with Budget Constraints and Available Space

Finding the perfect balance between system size, budget constraints, and available space is key to designing a solar power system that is both practical and efficient. Here are some tips to help you strike the right balance:

1. Prioritize energy efficiency: Before sizing your solar power system, consider investing in energy efficiency upgrades. By reducing your energy needs, you may be able to install a smaller and more cost-effective solar power system.
2. Assess your roof or available mounting space: The size and orientation of your roof or available ground space will determine the number of solar panels you can install. Take measurements and account for any obstructions, such as chimneys or shading, that may affect the system's performance.
3. Determine your budget: Understand the costs associated with purchasing and installing a solar power system, and set a realistic budget that aligns with your financial goals. Keep in mind that incentives, rebates, and tax credits can help offset the costs of solar power systems.
4. Consult with professionals: Reach out to solar power system installers or consultants to discuss your energy

needs, available space, and budget. They can help you design a system that meets your requirements while adhering to your financial constraints.

By carefully considering your energy needs, available space, and budget constraints, you can effectively size your solar power system to match your consumption patterns. This approach ensures that your solar power system will be both efficient and practical, setting you on the path to a sustainable and cost-effective energy future.

CONCLUSION: EMPOWERING YOUR SOLAR POWER JOURNEY WITH INFORMED CHOICES

How Understanding Energy Needs Streamlines the Solar Power System Design Process

A thorough grasp of your energy needs is the cornerstone of a successful solar power journey. By comprehending your consumption patterns, you'll be better equipped to streamline the design process for your solar power system. This understanding allows you to make well-informed decisions, ensuring your solar installation is tailored to your unique requirements. A solar power system designed with precision will not only be more efficient, but also enhance your return on investment while minimizing waste and excess generation.

The Benefits of a Customized Solar Power System for Your Unique Situation

A solar power system customized to your specific energy needs offers numerous advantages. By accurately assessing your consumption and integrating energy efficiency measures, you can enjoy the following benefits:

1. Optimal system performance: A tailored solar power system will operate at its peak efficiency, generating just the right amount of electricity to meet your needs without overloading or underutilizing your solar panels.
2. Financial savings: With a solar power system matched to your consumption, you can optimize your investment, reduce your reliance on grid electricity, and enjoy significant savings on your utility bills over time.
3. Environmental impact: By installing a solar power system that aligns with your energy needs, you contribute to reducing greenhouse gas emissions and fostering a greener, more sustainable future.
4. Enhanced self-sufficiency: A well-designed solar power system can provide you with a greater sense of energy independence, reducing your reliance on traditional energy sources and empowering you to take control of your energy consumption.

By understanding your energy needs and making informed choices, you're embarking on a solar power journey that will lead to a brighter, more sustainable future. With a customized solar power system tailored to your unique situation, you'll reap the rewards of your investment while contributing to a cleaner, greener world.

[CHAPTER 4] — LET THERE BE LIGHT: CHOOSING THE RIGHT SOLAR PANEL SYSTEM AND MOUNTING SOLUTIONS

INTRODUCTION: NAVIGATING THE WORLD OF SOLAR PANEL SYSTEMS AND MOUNTING SOLUTIONS

Welcome to Chapter 4, where we'll guide you through the process of choosing the right solar panel system and mounting solution for your needs.

Making informed decisions about these essential components of your solar power setup can significantly impact the performance, aesthetics, and long-term satisfaction you'll enjoy from your investment.

The Importance of Selecting the Right Solar Panel System

Selecting the right solar panel system is crucial to ensure that it meets your energy needs, budget, and space constraints. The type of solar panels you choose will affect your system's overall efficiency, aesthetics, and cost. Taking the time to compare different solar panel types and understand their unique characteristics will help you make the best decision for your specific situation.

Factors to Consider When Choosing Mounting Solutions

When choosing mounting solutions, several factors come into play. Your available space, the type of solar panels you have selected, and local building codes all influence the most suitable mounting option for your solar power system. In this chapter, we will explore the different mounting solutions, their pros and cons, and how to determine the best fit for your needs.

Understanding Solar Panel Types: Monocrystalline, Polycrystalline, and Thin-Film

Navigating the solar panel market can be overwhelming due to the variety of options available. However, solar panels can generally be grouped into three main categories: monocrystalline, polycrystalline, and thin-film. In this section, we will delve into the key differences between these solar panel types, focusing on efficiency, aesthetics, and cost to help you make an informed decision.

Comparing Efficiency, Aesthetics, and Cost

Monocrystalline solar panels are known for their high efficiency and sleek appearance. They are made from a single crystal structure, resulting in a uniform, dark color. Although they tend to be more expensive, their higher efficiency means they require less space to generate the same amount of power as other panel types.

Polycrystalline solar panels, on the other hand, are made from multiple crystal structures, giving them a distinctive blue hue. While they are generally less efficient than monocrystalline panels, they come with a lower price tag, making them a popular choice for budget-conscious consumers.

Thin-film solar panels are the most lightweight and flexible option, with a unique appearance that can be customized to blend seamlessly with various surfaces. Although they have the lowest efficiency of the three types, their lower cost and adaptability make them an attractive choice for certain applications, such as building-integrated photovoltaics (BIPV).

Determining the Best Fit for Your Needs

Choosing the right solar panel type for your needs involves considering your priorities in terms of efficiency, aesthetics, cost, and available space. If maximizing efficiency and having a sleek appearance are your top concerns, monocrystalline panels might be the best choice. If you're working with a tighter budget and are willing to sacrifice some efficiency, polycrystalline panels could be a suitable option. Lastly, if you need a flexible, lightweight solution or are interested in integrating solar panels into your building's structure, thin-film panels could be the way to go.

SOLAR PANEL MOUNTING SOLUTIONS: GROUND, ROOF, AND POLE

Selecting the right mounting solution for your solar panels is essential for optimizing your system's performance and ensuring its durability. In this section, we will explore the three primary mounting options: ground, roof, and pole mounts.

Pros and Cons of Each Mounting Type

Ground Mounts:

Pros:

- Easy access for maintenance and cleaning
- Optimal angle and orientation can be easily achieved
- No roof penetrations required, reducing potential for leaks
- Can be expanded easily if more panels are needed

Cons:

- Requires available land or yard space
- Can be more expensive than roof mounts due to additional materials required to create a structure and associated labor

Roof Mounts:

Pros:

- Utilizes existing roof space, no additional land needed
- Typically less expensive than ground mounts
- Less noticeable and more aesthetically pleasing

Cons:

- May require roof reinforcements or modifications
- Limited by roof location, size, shape, and orientation
- Access for maintenance and cleaning can be more difficult
- Potential shading from nearby structures or trees

Pole Mounts:

Pros:

- Allows for optimal panel angle and orientation
- Elevated off the ground, reducing shading and debris accumulation
- Can be installed in various locations

Cons:

- Requires additional materials and labor, making it more expensive than roof mounts
- May be less aesthetically pleasing
- May require additional permitting

Compatibility with Various Solar Panel Types

Most mounting solutions are designed to be compatible with various solar panel types, including monocrystalline, polycrystalline, and thin-film panels. However, it's essential to verify that your chosen mounting system is compatible with your specific panels and can support their weight and size.

Factors Affecting the Choice of Mounting Solution

Several factors should be considered when selecting the most suitable mounting solution for your solar power system:

1. Available Space: Assess your available land, yard, or roof space to determine which mounting option is the most feasible.
2. Shading: Evaluate the potential for shading from nearby structures, trees, or other obstacles that could impact your system's performance.
3. Aesthetics: Consider the visual impact of the mounting solution on your property and whether it aligns with your preferences or any homeowners' association rules.
4. Budget: Compare the costs of different mounting solutions, keeping in mind that ground and pole mounts may require additional materials and labor.

5. Local Regulations and Permitting: Check your local building codes and permitting requirements, as they may influence your choice of mounting solution.
6. Maintenance and Accessibility: Consider how easy it will be to access your solar panels for cleaning and maintenance with each mounting option.

By carefully evaluating these factors, you can choose the most appropriate mounting solution for your solar panel system, ensuring optimal performance and longevity.

SOLAR PANEL ORIENTATION AND TILT: MAXIMIZING ENERGY PRODUCTION

The performance of your solar power system is significantly influenced by the orientation and tilt of your solar panels. The ideal condition for maximum solar generation occurs when the panels are directly perpendicular to the sun's rays, ensuring they capture the most sunlight possible. By optimizing these factors, you can maximize the amount of energy your panels produce, increasing your system's overall efficiency and savings.

The Impact of Orientation and Tilt on Solar Panel Performance

Orientation refers to the direction your solar panels face, while tilt refers to the angle at which they are mounted relative to the horizontal. Both factors are critical in determining the amount of sunlight your panels receive and, consequently, the amount of energy they produce.

Orientation: Solar panels should be positioned to receive maximum sunlight exposure throughout the day. In the Northern Hemisphere, panels should face south to capture the most sunlight, while in the Southern Hemisphere, they should face north.

Tilt: The tilt of your solar panels affects the angle at which sunlight strikes the surface, influencing the panels' energy production. An optimal tilt angle allows the panels to capture the most sunlight possible during peak solar hours.

Optimal Angles and Directions for Your Location

The optimal orientation and tilt for your solar panels depend on your location's latitude and the specific characteristics of your installation site. Here are some general guidelines to follow:

Orientation: As mentioned earlier, panels should face south in the Northern Hemisphere and north in the Southern Hemisphere. However, if your roof or mounting location doesn't allow for optimal orientation, you can still achieve good results by installing panels within 30 degrees of the ideal direction.

Tilt: To determine the optimal tilt angle for your location, you can start by using your latitude as a baseline. For example, if your location's latitude is 40 degrees, you can begin with a 40-degree tilt angle. Adjustments can be made based on the time of year:

- In summer, decrease the tilt angle (making the panels more horizontal) by 10-15 degrees from baseline for better sun exposure.
- In winter, increase the tilt angle (making the panels more vertical) by 10-15 degrees from baseline to capture more sunlight when the sun is lower in the sky.

It's essential to consider factors such as shading from nearby structures or trees, roof angle, and local climate when determining the best orientation and tilt for your solar panels. In some cases, consulting with a solar professional can help you find the most effective configuration for your specific situation.

By optimizing the orientation and tilt of your solar panels, you can significantly improve your solar power system's performance, ensuring that you get the most out of your investment.

SPACE CONSIDERATIONS: HOW MUCH ROOM DO YOU NEED FOR YOUR SOLAR PANELS?

Assessing Your Available Space

Before beginning your project, it's essential to evaluate the space available for installing your solar panels. Whether you're considering a rooftop, ground, or pole-mounted system, take the time to measure the area where the panels will be installed.

Consider factors such as shading from trees, buildings, or other structures, as these can impact your system's efficiency. Also, account for any local building codes or homeowners' association regulations that may limit the size or placement of your solar panel system.

Calculating the Number of Panels Required

Once you have a clear understanding of your available space, it's time to calculate the number of solar panels you'll need. To determine this, start by reviewing your energy needs assessment from Chapter 3. Divide your daily energy consumption by the average output of a single solar panel.

Keep in mind that a solar panel's output depends on factors like its orientation and the average solar irradiance in your area. In simpler terms, solar irradiance indicates the energy a panel can produce under perfect conditions. For example, if your location has an average irradiance value of 3.5, it means in an average day your panel would generate the same amount of energy as it would if it were producing maximum power output for 3.5 hours. You can find your area's unique irradiance value online.

Keep in mind that it's generally better to slightly overestimate the number of panels needed to account for factors like shading or panel degradation over time.

Planning for Future Expansion

As you're evaluating your space and calculating the number of panels required, it's wise to plan for potential future expansion. Energy needs can change over time, especially if you plan to add more appliances, an electric vehicle, or any other energy-consuming devices to your household. By leaving room for additional panels, you'll be prepared to expand your solar power system as your needs evolve, ensuring that you continue to benefit from clean, renewable energy.

SIZING YOUR INVERTER AND BATTERY BANK: KEY CONSIDERATIONS FOR A BALANCED SOLAR POWER SYSTEM

An essential part of designing an efficient and reliable solar power system is selecting the right inverter and battery bank. In this section, we will discuss the key considerations for sizing these components to ensure optimal performance and energy independence.

Inverter Sizing

The inverter is responsible for converting the DC electricity generated by your solar panels into AC electricity, which can be used by your home appliances or fed back into the grid. When sizing the inverter, keep the following factors in mind:

Solar array output: Your inverter should be capable of handling the maximum output of your solar array and the maximum potential household demand. A general rule of thumb is to choose an inverter with a capacity of around 10-20% more than the peak expected power output of your solar panels.

Efficiency: Inverters have varying efficiency ratings. Higher efficiency inverters will convert more of the generated solar power into usable electricity, resulting in less energy loss.

Grid-tied vs. off-grid systems: Grid-tied systems may require specific types of inverters that comply with grid connection standards, while off-grid systems may require inverters with built-in charge controllers for battery management.

Battery Bank Sizing

A battery bank stores excess solar energy, allowing you to use it during times of low solar production, such as at night or during cloudy days. To size your battery bank, consider the following factors:

Energy storage capacity: Calculate the amount of energy you want to store based on your daily energy consumption and the desired number of backup days. The battery bank's capacity should be sufficient to cover your energy needs during periods of low solar production.

Depth of Discharge (DoD): The DoD refers to the percentage of a battery's capacity that can be safely discharged before it needs to be recharged. To prolong battery life, choose a battery bank with a higher DoD and size it accordingly.

Battery type: Different battery types, such as lead-acid and lithium-ion, have varying characteristics, lifespans, and costs. Consider the advantages and disadvantages of each battery type when sizing your battery bank.

By carefully considering the sizing of your inverter and battery bank, you can create a balanced solar power system that meets your energy needs while optimizing efficiency and performance. Remember to consult with solar professionals to ensure that your system is designed and installed correctly so you can maximize the benefits of your investment.

MOUNTING HARDWARE AND ACCESSORIES: ENSURING A SECURE AND EFFICIENT INSTALLATION

Choosing the Right Mounting Hardware

Selecting appropriate mounting hardware is crucial for a safe, secure, and efficient solar panel installation. The type of hardware you'll need depends on your chosen mounting solution, whether it's ground, roof, or pole-mounted. The materials used should be weather-resistant, durable, and compatible with your solar panels and mounting structure. Some common types of mounting hardware include:

Racking systems: These are the frameworks that support your solar panels and secure them in place. They should be made of sturdy materials like aluminum or stainless steel for long-lasting performance.

Clamps and brackets: These components connect your solar panels to the racking system, keeping them secure and properly aligned. Ensure they are compatible with your specific solar panel type.

Rails and splices: Rails provide additional support and stability for your solar panels, while splices join rails together for an extended mounting structure.

Essential Accessories for a Successful Installation

In addition to mounting hardware, there are several accessories that can enhance the performance and longevity of your solar power system. Some essential accessories for a successful installation include:

Grounding equipment: Proper grounding is essential for electrical safety and protects your solar power system from damage caused by lightning strikes or power surges.

Wire management clips: These clips help to organize and secure the wiring of your solar power system, reducing the risk of damage from exposure to weather or wildlife.

Safety equipment: Personal protective equipment (PPE) such as gloves, safety glasses, and harnesses are vital for a safe installation process, especially when working on roofs or at heights.

Monitoring systems: A solar monitoring system allows you to track your solar power system's performance and identify any potential issues early on, ensuring efficient operation.

By selecting the appropriate mounting hardware and accessories, you can optimize your solar power system's performance and ensure a safe, secure, and long-lasting installation.

BUILDING CODES, PERMITS, AND REGULATIONS: PLAYING BY THE RULES

Understanding Local Building Codes and Permit Requirements

Before embarking on your solar panel installation project, it's essential to familiarize yourself with local building codes and permit requirements.

Building codes are a set of guidelines and standards that ensure the safety and functionality of construction projects, including solar panel installations.

Permit requirements can vary depending on your location, so it's important to research the specific rules and regulations in your area. This may involve contacting your local building department, the authority having jurisdiction or consulting with a professional solar installer.

Navigating the Permitting Process

The permitting process for solar panel installations can be complex and time-consuming, but it's a necessary step to ensure your project complies with local regulations. To navigate this process, follow these steps:

Gather necessary documentation: This may include site plans, electrical diagrams, and structural engineering reports, depending on your local requirements.

Submit permit applications: Complete the required permit application forms and submit them to your local building department, along with any supporting documentation.

Schedule inspections: Once your permit is approved, you'll need to schedule inspections during and after the installation to verify that the project meets all code requirements.

Obtain final approval: After your solar panel system passes inspection, you'll receive final approval and permission to connect your system to the grid.

Staying Compliant with Regulations

Once your solar power system is up and running, it's important to stay compliant with local regulations. This may involve regular maintenance, reporting energy production to your utility company, or meeting specific safety standards. Keep up-to-date with any changes in regulations to ensure your system remains compliant and operates efficiently.

By understanding and adhering to local building codes, permits, and regulations, you can ensure a smooth and successful solar panel installation process. This compliance not only guarantees the safety and performance of your solar power system but also helps to maintain a positive relationship with local authorities and neighbors.

CONCLUSION: MAKING INFORMED DECISIONS FOR AN OPTIMAL SOLAR EXPERIENCE

As we conclude our exploration of solar panel systems and mounting solutions, it's essential to recognize the impact that informed decision-making can have on your solar power experience. By carefully considering the factors we've discussed throughout this chapter, you can confidently select the solar panel system and mounting solution that best align with your unique needs and preferences.

The right combination of solar panels and mounting options will not only enhance your solar experience but also maximize the performance and longevity of your system. By understanding the nuances of solar panel types, the pros and cons of various mounting solutions, and the importance of orientation and tilt, you can make choices that optimize energy production and ensure a satisfying return on your investment.

Equally important is staying informed about local building codes, permits, and regulations. By navigating the permitting process and ensuring compliance with all relevant rules, you can minimize potential roadblocks and maintain a positive relationship with local authorities and your community.

In summary, investing time and effort in understanding your options and making well-informed decisions will ultimately lead to a more satisfying solar power experience. By selecting the right solar panel system and mounting solution, you're laying the foundation for a bright, sustainable, and energy-efficient future.

[CHAPTER 5] — SAFETY FIRST: INSTALLING YOUR SOLAR POWER SYSTEM WITHOUT BREAKING A SWEAT

INTRODUCTION: THE IMPORTANCE OF SAFETY IN SOLAR POWER SYSTEM INSTALLATION

Installing a solar power system is an exciting and rewarding endeavor that can lead to significant long-term benefits, both environmentally and financially. However, it is essential to remember that safety must be the top priority throughout the entire installation process. By mitigating risks and preventing accidents, you can ensure a successful and trouble-free solar power system installation.

A well-executed installation plan incorporates safety measures at every stage, from pre-installation preparations to post-installation maintenance. This involves identifying potential hazards, using appropriate personal protective equipment (PPE), and following best practices when handling electrical components.

Taking the time to understand and implement safety measures is not only crucial for the well-being of the installation team but also guarantees the optimal performance and longevity of your solar power system.

In this chapter, we will explore various aspects of safety during the solar power system installation process, providing you with the knowledge and guidance necessary to confidently and safely embark on your solar journey.

PRE-INSTALLATION SAFETY MEASURES: LAYING THE GROUNDWORK FOR A SAFE INSTALLATION

Before you begin the actual installation of your solar power system, taking appropriate pre-installation safety measures is critical in ensuring a smooth and secure process. In this section, we will discuss the essential steps to lay the groundwork for a safe installation, from assessing your site for potential hazards to selecting the right personal protective equipment and assembling a well-prepared installation team.

Assessing Your Site for Potential Hazards

Before starting the installation process, it is essential to conduct a thorough site assessment to identify any potential hazards that could pose risks during the installation. This may include evaluating the stability of your roof, identifying obstructions or trip hazards, and checking for the presence of any hazardous materials. Additionally, be mindful of the weather conditions on the day of installation, as wet or windy conditions can significantly increase the risk of accidents.

Selecting Appropriate Personal Protective Equipment (PPE)

Choosing the right personal protective equipment (PPE) is crucial to ensuring the safety of your installation team. PPE can help protect against various hazards, such as electrical shocks, falls, and exposure to hazardous materials.

Key items to consider include safety goggles, gloves, hard hats, and non-conductive footwear. When selecting PPE, make sure to consult the manufacturer's guidelines and choose products that meet industry standards for quality and safety.

Assembling a Well-Organized and Prepared Installation Team

Having a well-prepared and organized installation team is vital to the overall safety and success of your solar power system installation. Ensure that all team members are adequately trained in the handling of solar equipment, understand the importance of safety procedures, and are familiar with the specific requirements of your installation site. Additionally, designate a team leader responsible for overseeing the project and ensuring that all safety protocols are followed. By fostering clear communication and a strong safety culture, you can significantly reduce the risk of accidents and ensure a smooth installation process.

SOLAR PANEL INSTALLATION: STEPS TO SAFELY SECURE YOUR PANELS

As you proceed with the installation of your solar panels, following a clear and well-structured plan will ensure that the process is both safe and efficient. This section will guide you through the necessary steps to safely secure your panels while adhering to best practices.

Verifying the Structural Integrity of Your Mounting Surface

Before you begin installing the solar panels, it's crucial to verify that the mounting surface—whether it's a roof, ground, or pole mount—can support the weight of the panels and mounting hardware. Consult a structural engineer or qualified professional to assess the load-bearing capacity of your chosen surface, and ensure that it meets or exceeds the requirements.

Safely Attaching Mounting Hardware and Securing Solar Panels

Once you have confirmed that your mounting surface is structurally sound, it's time to attach the mounting hardware. Follow the manufacturer's instructions and use the appropriate tools to ensure that the hardware is securely fastened. Remember to use the personal protective equipment (PPE) you selected earlier, such as safety gloves and goggles, to minimize the risk of accidents.

When installing the solar panels, work methodically and with care. Lift the panels with the help of another person to avoid strain injuries or equipment damage and secure them to the mounting hardware according to the manufacturer's guidelines. Be mindful of your surroundings and ensure that all team members communicate effectively to avoid accidents.

Properly Connecting Electrical Components and Wiring

After the solar panels are securely in place, it's time to connect the electrical components and wiring. Electrical safety is paramount during this stage, so make sure to adhere to local electrical codes and regulations.

> WARNING
>
> It's important to use a licensed electrician for power grid connection to ensure personal safety, the safety of your home and to comply with local building codes.

Keep in mind that if you are planning a grid-tied solar power system, a licensed electrician is required for final connections. Among their various responsibilities, the electrician will ensure that the main power supply to your home is properly disconnected during the installation process to minimize the risk of electrocution.

When connecting the electrical components, follow the solar power system's wiring diagram and the manufacturer's instructions. Use proper wire management techniques to ensure a neat and organized installation, which will reduce the risk of electrical issues and make maintenance easier in the future. After completing the wiring, double-check all connections before turning the power supply back on.

INVERTER AND BATTERY INSTALLATION: POWER CONVERSION AND STORAGE SAFETY

Selecting the Ideal Location for Your Inverter and Battery Bank

Choosing the right location for your inverter and battery bank is crucial for both safety and efficiency. Consider factors such as accessibility, protection from the elements, and proximity to your solar panels and main electrical panel. It's important to keep these components away from sources of excess heat, moisture, or any potential fire hazards.

Following Safety Guidelines for Electrical Connections When installing the inverter and battery bank, it's essential to adhere to safety guidelines for electrical connections. Ensure that all wiring is appropriately sized for the expected current and voltage levels and apply proper grounding techniques. Always follow the manufacturer's instructions and local electrical codes to prevent electrical hazards.

Ensuring Proper Ventilation and Temperature Control

Both inverters and battery banks require proper ventilation and temperature control to operate efficiently and safely. Overheating can lead to reduced performance, shortened component lifespan, or even pose a fire risk. Make sure that the installation location provides adequate airflow, and if necessary, install additional ventilation or cooling systems to maintain an optimal operating temperature. Regularly inspect and maintain these systems to ensure continued safety and efficiency.

ELECTRICAL SAFETY: PROTECTING YOUR SOLAR POWER SYSTEM AND HOME

Installing Grounding and Surge Protection Systems

To protect your solar power system and home from potential electrical hazards, it's crucial to install proper grounding and surge protection systems. Grounding ensures that electrical faults are safely channeled to the earth, reducing the risk of electrocution or fire. Surge protection systems guard your solar power system and home appliances against damage from power surges, which can occur due to lightning strikes or fluctuations in grid voltage.

To comply with basic grounding requirements for solar power systems, it's important to familiarize yourself with a few key concepts.

First, ensure that your system has an effective grounding electrode system, which may include a combination of ground rods, metallic water pipes, and building steel.

Next, make sure that all metal components of your solar power system, including mounting rails, frames, and enclosures, are properly bonded to this grounding electrode system.

Additionally, electrical grounding conductors should be appropriately sized and connected to minimize the risk of voltage fluctuations and potential hazards.

If you are unsure how to properly size or install a grounding system, it's critical to contact a qualified electrician or other qualified professional for support.

Fuses and Circuit Breakers: Essential Safety Measures in Your Solar Power System

Introduction to Fuses and Circuit Breakers

Fuses and circuit breakers are critical components in any electrical system, including your solar power system. They provide essential safety measures, protecting your system from overcurrent, short-circuit, and other electrical faults that could cause damage to your equipment or even pose a risk of fire. In this section, we'll discuss where to add fuses and circuit breakers in your solar power system and briefly explain how they work.

Where to Add Fuses and Circuit Breakers in Your Solar Power System

It's crucial to install fuses and circuit breakers at strategic points in your solar power system to ensure maximum protection. Key locations for these safety devices include:

- Between the solar panels and the charge controller: This prevents damage to the charge controller in case of a short circuit or overcurrent from the solar panels.
- Between the charge controller and the battery bank: This protects the battery bank from overcharging and potential damage caused by excessive current flow.
- Between the battery bank and the inverter:

This safeguards the inverter from any surge in current that could arise from the battery bank.

- On the AC output side of the inverter:
  This protects your household appliances and electrical system from any faults originating from the inverter or the solar power system.

How Fuses and Circuit Breakers Work

Fuses and circuit breakers both serve the same purpose — to protect electrical circuits by interrupting the flow of current in case of an overload or short circuit. However, they function differently.

Fuses: A fuse contains a thin metal wire or strip that melts when exposed to excessive current. When the metal melts, it breaks the electrical connection, effectively stopping the flow of current and protecting the rest of the system. Fuses are single-use devices, meaning once they have "blown," they need to be replaced with a new fuse of the correct rating.

Circuit Breakers: A circuit breaker is a reusable device that automatically trips or switches off when it detects an overcurrent or short circuit. This interruption in the flow of current protects the electrical circuit from damage. After tripping, a circuit breaker can be manually reset, allowing it to be used repeatedly without the need for replacement. Circuit breakers offer a more convenient and cost-effective solution for protecting your solar power system.

By installing fuses and circuit breakers in your solar power system, you're taking essential steps to ensure the safety and longevity of your investment. These devices help prevent damage to your equipment and minimize the risk of electrical hazards, giving you peace of mind as you enjoy the benefits of your solar energy system.

SIZING ELECTRICAL CABLES: CHOOSING THE RIGHT SIZE FOR YOUR SOLAR POWER SYSTEM

Introduction to the Art of Cable Sizing

Selecting appropriately sized electrical cables is crucial to the efficiency, safety, and reliability of your solar power system. The right size ensures minimal power loss, reduces the risk of overheating, and prevents voltage drops that can impact your system's performance. It's essential to understand the importance of cable sizing and how to make the right selection for your solar power system.

The Significance of Appropriate Cable Selection

Choosing the correct cable size is critical for several reasons:

Efficiency: Properly sized cables help minimize energy loss due to resistance, ensuring maximum power transfer from your solar panels to your battery bank and inverter.

Safety: Undersized cables can overheat, potentially causing damage to your equipment or even posing a fire risk.

System Performance: Correctly sized cables prevent voltage drops, ensuring your equipment operates at optimal levels and delivers the expected power output.

A Step-by-Step Guide to Sizing Cables for Your Solar Installation

To size the electrical cables for your solar power system, follow these steps:

Determine the Current: Calculate the maximum current that will flow through the cable, based on the power output of your solar panels, charge controller, and inverter. Remember to consider both the DC and AC sides of your system.

Determining the Current for Your Solar Power System

To determine the current for your solar power system, you'll need to consider the system's components and their individual specifications. Follow these steps to calculate the current:

1. Review solar panel specifications: Check the solar panel manufacturer's datasheet for the maximum power

current (Imp) and the short-circuit current (Isc). Both values are given in amperes (A).

2. Calculate solar panel string current: If you have multiple solar panels connected in series, the current remains the same as for a single panel. However, if you connect panels in parallel, the current increases. To find the total current for parallel connections, add the currents of each panel together.

For example, if you have three panels connected in parallel, each with an Imp of 5A:

Total current (parallel) = Imp1 + Imp2 + Imp3 = 5A + 5A + 5A = 15A

3. Review charge controller and inverter specifications: Consult the manufacturer's datasheets for your charge controller and inverter. Look for the maximum input current and output current ratings. These values will help you determine the appropriate cable size for the connections between the solar panels, charge controller, battery bank, and inverter.
4. Estimate the load current: To determine the load current, divide the total wattage of your electrical loads by the system voltage. This will give you an estimate of the current drawn by your loads.

For example, if you have a 2,000W load and a 48V system:

Load current = Total wattage / System voltage = 2,000W / 48V = 41.67A

5. Factor in safety margins: To ensure safe operation, it's a good practice to include a safety margin when determining current. A common safety margin is 25%. To apply this safety margin, multiply the calculated current by 1.25.

Using the previous example:

Safety margin applied = Load current * 1.25 = 41.67A * 1.25 = 52.09A

So, in this example you would select a cable that is capable of carrying more than 52 Amps.

By determining the current for your solar power system, you'll be able to select the appropriate cable sizes to ensure optimal performance, safety, and efficiency.

Choose the Appropriate Wire Material: Copper and aluminum are the most commonly used materials for electrical cables. Copper generally offers better conductivity and is more resistant to corrosion, but it's also more expensive than aluminum. Your choice will depend on your budget and system requirements.

Calculate the Cable Size: Use the calculated current and the allowable voltage drop to determine the appropriate cable size. The ultimate authority for determining the correct cable size is the National Electrical Code (NEC), though various online calculators and cable sizing charts are available to help with this step. Be sure to consider factors such as cable length, material, and ambient temperature, as these can affect cable sizing.

Verify the Cable Rating: Ensure the chosen cable size can handle the maximum current and voltage your system will generate. *Never let your cable become the fuse!* It's also essential to consider the cable's temperature rating, as excessive heat can reduce the cable's capacity.

Consult the National Electrical Code for properly sizing of electrical conductors. A free version is available to review online through the National Fire Protection Association.

By taking the time to size your electrical cables correctly, you'll optimize your solar power system's efficiency and safety. Proper cable sizing ensures your equipment operates at its best, maximizing the return on your solar energy investment.

Understanding and Implementing Safety Disconnects

Safety disconnects play a vital role in protecting both your solar power system and the people who interact with it. These devices allow you to quickly and safely interrupt the flow of electricity in case of an emergency, maintenance, or system inspection. Familiarize yourself with the types and locations of safety disconnects in your solar power system, and ensure they are easily accessible and properly labeled.

Properly Labeling and Organizing Wiring for Ease of Maintenance

A well-organized and labeled wiring system not only makes your solar power system installation look professional but also enhances safety and simplifies maintenance. Properly label all wiring, connectors, and components according to their function and voltage levels. Keep wiring neat and secure, using cable management systems such as cable ties or conduit, to prevent damage or accidental disconnection. By maintaining an organized and clearly labeled wiring system, you'll make future inspections, troubleshooting, and maintenance tasks safer and more efficient.

POST-INSTALLATION SAFETY: ONGOING MAINTENANCE AND MONITORING

Maintaining the safety and efficiency of your solar power system is an ongoing task. After installation, it's crucial to establish a comprehensive routine that includes regular inspections, maintenance, and monitoring to ensure your system continues to perform optimally while minimizing risks.

Establishing a Routine Inspection and Maintenance Schedule

To keep your solar power system running smoothly and safely, it's essential to create a regular inspection and maintenance schedule.

This should include visual inspections for signs of wear and tear, corrosion, and loose connections, as well as cleaning solar panels to remove dirt, dust, and debris that can reduce their efficiency.

Additionally, be sure to check the mounting system, electrical components, and battery bank for any potential issues. Following a consistent schedule helps to identify and address any problems early, reducing the likelihood of safety hazards or system failures.

Identifying and Addressing Potential Safety Hazards

As part of your ongoing maintenance routine, be vigilant in identifying and addressing any potential safety hazards. This includes inspecting wiring for damage or signs of overheating, ensuring that all connections are secure, and verifying that grounding and surge protection systems are functioning properly.

If you discover any issues, it's crucial to address them safely and promptly to minimize risks and maintain the safety and performance of your solar power system.

Implementing a Remote Monitoring System for Real-Time System Performance Tracking

A remote monitoring system can be a valuable tool in maintaining the safety and efficiency of your solar power system. By providing real-time data on system performance, it allows you to quickly identify any issues that may arise, such as drops in energy production or irregularities in voltage or current levels.

In addition, many remote monitoring systems can alert you to potential safety hazards, enabling you to take immediate action to resolve the issue. By staying informed and proactive, you can ensure the ongoing safety and performance of your solar power system.

CONCLUSION: HARNESSING SOLAR POWER SAFELY AND EFFECTIVELY

Throughout this chapter, we've emphasized the importance of safety when installing and maintaining a solar power system. By following the guidelines and recommendations provided, you can mitigate risks and prevent accidents, ensuring a successful and trouble-free installation. As you move forward with your solar power journey, remember that taking the time to prioritize safety is not only essential for the well-being of you and your family, but it also contributes to the longevity and performance of your solar power system.

Now that you've gained the knowledge and confidence to approach solar power system installation safely, it's time to move on to the next chapter. In Chapter 6, *Plug and Play: Connecting Your System In 5 Easy Steps*, we'll guide you through the process of connecting your solar power system to your home. With a clear, step-by-step approach, you'll be ready to enjoy the benefits of clean, renewable solar energy in no time.

[CHAPTER 6] — PLUG AND PLAY: CONNECTING YOUR SYSTEM IN 5 EASY STEPS

INTRODUCTION: BRINGING YOUR SOLAR POWER SYSTEM TO LIFE

The moment has finally arrived: after all the planning, researching, and preparation, it's time to bring your solar power system to life. The significance of proper system connections cannot be overstated. Correctly connecting your solar panels, charge controller, battery bank, and inverter ensures that your system will function efficiently and safely, allowing you to harness the sun's power to its fullest potential.

Ensuring a smooth and hassle-free setup requires attention to detail and a step-by-step approach. This chapter will guide you through the process, making sure you're well-equipped to connect your system with confidence.

These instructions assume you have a battery backup for your system. Considerations for connecting to grid power are also discussed where appropriate.

As you work through each step, remember that the time and effort you invest now to understand the process will pay off in the long run, as you enjoy the benefits of clean, renewable energy generated by your very own solar power system. So, let's dive in and start connecting your solar power system, one component at a time.

STEP 1: PREPARING FOR GRID CONNECTION

The main purpose of this book is to assist readers in learning about DIY solar setups. However, Step 1 is included for those who plan on having a grid-tied solar system. Standalone DIY solar installations, on the other hand, won't be connected to your utility grid. If that's the case for you, continue to Step 2.

If you do intend to have a grid-tied system, seek the assistance of a professional and ensure final connections are coordinated with your utility and performed by a qualified professional. Also, note that a battery bank may not be necessary. If you're uncertain about your abilities, it's essential to seek professional advice to avoid any potential danger or regulatory issues.

If you are connecting your solar power system to the grid, you first need to lay the groundwork by coordinating with your utility company, obtaining necessary permits, and meeting grid interconnection requirements. Proper preparation ensures a smooth process and helps avoid unexpected roadblocks during system integration.

Coordinating with Your Utility Company

Begin by contacting your utility company to inform them about your plans to install a solar power system. They will provide you with important information about their specific grid connection requirements, timelines, and fees. Establishing a clear line of communication with your utility company early in the process will help streamline the connection process and avoid potential delays.

Obtaining Necessary Permits and Meeting Grid Interconnection Requirements

In most areas, you will need to obtain permits before connecting your solar power system to the grid. These permits may include electrical, building, and zoning permits. Check with your local government offices and utility company to determine the required permits and their associated fees. Additionally, your utility company may have specific grid interconnection requirements, such as safety devices or system design standards, that you must meet to connect your system.

Installing a Bi-Directional Meter for Net Metering

Net metering allows you to receive credit for any excess energy your solar power system generates and feeds back into the grid. To participate in net metering, you'll need a participating utility and a bi-directional meter installed at your home. This meter measures both the electricity your home consumes from the grid and the surplus electricity your solar system generates.

Contact your utility company to schedule the installation of a bi-directional meter. They will either install a new meter or modify your existing one, depending on their policies and the type of meter currently in use.

By properly preparing for grid connection, you set the stage for a successful solar power system integration. With your utility company informed and permits in hand, you're one step closer to harnessing the sun's power and enjoying the benefits of clean, renewable energy.

STEP 2: CONNECTING SOLAR PANELS TO THE CHARGE CONTROLLER

Verifying Proper Solar Panel Wiring

Before connecting your solar panels to the charge controller, it's crucial to verify that your solar panel wiring is done correctly. Improper wiring can lead to reduced system efficiency, potential damage to components, and even safety hazards.

Start by checking the solar panel connections. Ensure that the connectors are clean, free of dirt or corrosion, and securely fastened. Inspect the wiring for any signs of damage, such as frayed insulation or exposed wires.

Make sure that the positive (+) and negative (-) wires from the solar panels are correctly identified and connected, as reversing the polarity can cause damage to the charge controller or other system components.

If your solar panels are wired in series or parallel, double-check that they are configured according to your system design.

Safely Connecting Solar Panel Strings to the Charge Controller

Once you have verified the proper wiring of your solar panels, it's time to connect them to the charge controller. The charge controller plays a vital role in managing the flow of electricity from the solar panels to the battery bank, ensuring optimal charging and preventing overcharging or discharging.

Before you begin, make sure to consult the user manual and fully comprehend the connection instructions and any associated warnings. Ensure that the charge controller is turned off or disconnected from the battery bank before making any connections. This is necessary to prevent electrical hazards or potential damage to the controller during the process.

Next, locate the input terminals on the charge controller marked for solar panel connections. These terminals are typically labeled as PV+, PV-, or similar designations. Connect the positive (+) wire from your solar panel string to the positive PV+ terminal and the negative (-) wire to the negative PV- terminal. Make sure the connections are secure and that there is no risk of accidental disconnection or short-circuiting.

Once you've connected the solar panels to the charge controller, double-check the connections to ensure they are correct and secure. After verifying the connections, you can proceed with connecting the charge controller to the battery bank and inverter, as outlined in the next steps.

STEP 3: INTEGRATING THE BATTERY BANK INTO YOUR SYSTEM

Safely Connecting the Charge Controller to the Batteries

Once you have chosen an appropriate location for your battery bank (see Chapter 5), it's time to connect the charge controller. Begin by consulting equipment manuals to fully understand instructions and warnings. Turn off the charge controller to avoid any electrical hazards or potential damage during the connection process.

Locate the output terminals on the charge controller marked for battery connections, typically labeled as BAT+, BAT-, or similar designations. Connect the positive (+) wire from the charge controller to the positive terminal on your battery bank, and the negative (-) wire to the negative terminal. Ensure that the connections are secure and that there is no risk of accidental disconnection or short-circuiting.

Ensuring Proper Battery Configuration and Polarity

Before finalizing the battery connections, double-check the battery configuration to ensure it is aligned with your system design. If your batteries are connected in series, parallel, or a combination of both, verify that the connections are correct and secure.

Confirm that the polarity of the battery connections is accurate, with the positive (+) wire connected to the positive terminal and the negative (-) wire connected to the negative terminal. Reversing the polarity can cause damage to the charge controller, inverter, or other system components.

After you have ensured the proper battery configuration and polarity, you can proceed with connecting the battery bank to the inverter in the next step.

STEP 4: ESTABLISHING THE INVERTER CONNECTION FOR POWER CONVERSION

The Crucial Role of the Inverter in Your Solar Power System

The inverter is a vital component in your solar power system, as it's responsible for converting the DC power generated by your solar panels and stored in the battery bank into AC power that can be used by your household appliances.

Establishing a proper connection between your solar panel system, battery bank, and the inverter ensures smooth and efficient energy conversion, enabling you to reap the full benefits of your solar energy investment.

Preparing for a Successful Inverter Connection

To make the inverter connection process as smooth and hassle-free as possible, it's essential to understand the different settings and options available in the inverter, as well as the requirements of your solar panel system and grid connection (if applicable). This knowledge will enable you to configure the inverter correctly, avoid potential issues, and optimize your solar power system's performance. In this section, we'll guide you through the process of understanding and configuring your inverter to ensure a successful power conversion.

Understanding Inverter Settings and Options

It's important to familiarize yourself with the various settings and options available on your inverter. Inverters typically offer settings related to voltage, frequency, and grid type, among others. Consult the manufacturer's manual or guidelines to understand the purpose and function of each setting.

Matching Inverter Settings to Your Solar Panel System and Grid Requirements (if Grid Connected)

Once you're familiar with the available settings and options, it's crucial to fine-tune your inverter to align with your solar panel system's specifications, and, if you're connecting to the grid, the requirements set by your utility company.

Adjust the voltage and frequency settings according to your solar panel system's output and the grid's operating parameters, if applicable. Enable any required protection features, such as anti-islanding, to ensure safe operation and compliance with utility company regulations.

Anti-Islanding: What is it and why do we Care?

Anti-islanding is a safety feature implemented in grid-tied inverters to protect both the solar power system and the electrical grid. It is designed to detect when the grid has lost power due to an outage, and subsequently disconnect the solar power system from the grid to prevent feeding power back into it. This is important for the safety of utility workers who may be repairing the grid, as well as for preventing damage to your solar power system.

During a grid outage, if a solar power system continues to feed electricity into the grid, it creates an "island" of powered electrical circuits, which can pose a risk to utility workers and others nearby who might assume the power lines are de-energized. By disconnecting the solar power system from the grid, anti-islanding protection ensures that the electrical circuits are not energized when they should not be.

Anti-islanding protection typically works by continuously monitoring the grid's voltage and frequency. If the inverter detects an abnormality, such as a sudden drop in voltage or frequency, it interprets this as a grid outage and quickly disconnects the solar power system. Once the grid has been restored and the inverter detects stable voltage and frequency, it automatically reconnects the solar power system to the grid.

Testing the Inverter to Ensure Proper Function

After configuring and your inverter, it's important to test its functionality to verify that it's operating correctly and safely. Follow the user manual's testing procedure, which may include checking for proper voltage and frequency output, testing the anti-islanding protection, and ensuring that the inverter's display and monitoring functions are working correctly. If you encounter any issues during testing, consult the user manual or contact the inverter manufacturer for troubleshooting assistance.

Properly configuring your inverter is a critical step in setting up your solar power system. By taking the time to understand your inverter's settings, matching them to your solar panel system and grid requirements, and testing its functionality, you help ensure optimal performance and safe operation for your solar power system.

STEP 5: FINALIZING CONNECTIONS AND VERIFYING SYSTEM PERFORMANCE

Linking the Battery Bank to the Inverter Completing the Circuit

Understanding the Connection Process

To connect the battery bank to the inverter, you'll need to follow the manufacturer's guidelines and ensure you use the appropriate cables and connectors. It's crucial to understand the connection process and the role each component plays in the system. The battery bank stores the DC electricity generated by your solar panels, and the inverter converts this DC power into AC power that your home appliances can use.

Selecting the Right Cables and Connectors

Throughout your system, it's critical that you choose cables and connectors specifically designed for the voltage and current ratings of the equipment you're connecting, in this case, your battery bank and inverter. Please refer to Chapter 5 for a detailed discussion on cable sizing. Using the wrong cables can result in increased resistance, voltage drops, and even potential fire hazards. Make sure to use cables of the appropriate size (gauge) and type, as well as connectors that match your battery and inverter terminals.

Ensuring Correct Wiring and Polarity

When connecting the battery bank to the inverter, it's crucial to ensure that the wiring and polarity are correct. Connect the positive (+) terminal of the battery bank to the positive (+) input terminal of the inverter and the negative (-) terminal of the battery bank to the negative (-) input terminal of the inverter. This step helps prevent damage to the components and ensures proper functioning of the system.

Verifying Connections and Testing the System

After making the connections between the battery bank and the inverter, double-check to ensure everything is secure and correctly connected. It's a good practice to test the system by switching on a small load, like a light or a small appliance, to ensure the inverter is functioning correctly and supplying AC power.

Revisiting Grounding and Surge Protection:

Double-Checking Your Setup Now that your system is connected, it's essential to review the grounding and surge protection measures you implemented in Chapter 5. Make sure that all connections are secure and that the grounding and surge protection systems are correctly installed, providing optimal safety for your solar power system.

Assessing System Performance and Safety Features:

The Final Check Before you declare your solar power system fully operational, take the time to verify its performance and safety features. Perform all functional testing prescribed by your systems user's manuals. Check that all safety disconnects are functional and test the anti-islanding protection if you're connected to the grid. Monitor the system's output to ensure it matches the expected performance based on your solar panel system and location. Once you're confident that everything is working correctly and safely, you can enjoy the benefits of your solar power system with peace of mind.

CONCLUSION: POWERING YOUR WORLD WITH SOLAR ENERGY

As we come to the end of Chapter 6, you should now have a solid understanding of the steps involved in connecting your solar power system to ensure a seamless, efficient, and safe operation. From preparing for a possible grid connection and incorporating charge controllers to configuring inverters and establishing appropriate electrical connections, each step plays a critical role in harnessing the power of the sun to meet your energy needs.

Remember, it is essential to adhere to local regulations and best electrical engineering practices when setting up your solar power system. A well-connected system not only maximizes efficiency and performance but also ensures the safety and longevity of your investment.

As you embark on this exciting journey to a greener and more sustainable future, it's important to remember that the work doesn't end with the installation and connection of your solar power system.

In the next chapter, we'll dive into the world of maintenance and monitoring, guiding you through the necessary steps to keep your system in optimal condition and shining bright. From routine inspections and cleaning to monitoring system performance, we'll cover everything you need to know to ensure your solar power system continues to be a reliable and efficient source of energy for years to come. So, let's keep it shining and continue on to Chapter 7!

[CHAPTER 7] — KEEP IT SHINING: MAINTAINING AND MONITORING YOUR SOLAR POWER SYSTEM

INTRODUCTION: THE IMPORTANCE OF REGULAR MAINTENANCE AND MONITORING

Congratulations on taking the leap into the world of renewable energy with your freshly installed solar power system! However, the journey doesn't end with the installation. Just like any valuable investment, your solar power system requires regular maintenance and monitoring to ensure optimal performance, longevity, and a seamless solar experience. By taking the time to care for your solar power system, you'll be better equipped to detect and resolve potential issues early on, avoiding costly repairs and downtime.

In this chapter, we'll delve into the ins and outs of maintaining and monitoring each component of your solar power system, from the solar panels themselves to the battery bank, inverter, and charge controller. We'll also explore the importance of a properly maintained electrical system and discuss how to utilize remote monitoring tools to keep a watchful eye on your system's performance.

So, strap in and prepare to embark on the road to solar power system mastery! With the right knowledge, tools, and proactive approach, you'll be well on your way to getting the most out of your investment and enjoying the benefits of clean, renewable energy for years to come.

SOLAR PANEL MAINTENANCE: KEEPING YOUR PANELS CLEAN AND EFFICIENT

Routine Inspection and Cleaning Techniques

Regular inspection and cleaning of your solar panels are essential to ensure their peak performance and efficiency. Accumulation of dirt, dust, and debris can significantly reduce the amount of sunlight absorbed by the panels, decreasing their energy production. To maintain optimum efficiency, it's recommended that you inspect your solar panels every six months and clean them as needed.

When cleaning your solar panels, use a soft brush or sponge with water to gently remove dirt and debris. Avoid using harsh chemicals or abrasive materials that could damage the panels' surface. For hard-to-reach panels or those installed at a significant height, consider hiring a professional cleaning service to safely and effectively clean your solar array.

Addressing Common Solar Panel Issues

Apart from dirt and debris, solar panels may face other issues that can negatively affect their performance. Regular inspections can help you identify and address these problems early on. Some common issues to watch for include:

Damaged or broken panels: Check for cracks, chips, or other physical damages to the solar panel's surface or frame.

Loose connections: Inspect the wiring and connectors between your solar panels and the rest of the system, ensuring they're securely fastened and free from corrosion.

Pest infestation: Birds, rodents, or insects can nest under or around solar panels, potentially damaging the panels or their wiring. Regularly inspect for signs of infestation and take appropriate measures to deter pests.

The Impact of Weather and Environment on Solar Panel Maintenance

Weather and environmental factors play a significant role in the frequency and type of maintenance required for your solar panels. For instance, in areas with frequent rain or snowfall, your panels may need more frequent cleaning to remove accumulated debris. Additionally, extreme weather conditions like hail or strong winds can cause physical damage to the panels, warranting prompt repairs or replacements.

In areas with high humidity or coastal environments, salt and moisture in the air can cause corrosion of metal components and connectors. Regularly inspect these components for signs of corrosion and promptly address any issues to avoid long-term damage.

By staying vigilant and proactive in maintaining your solar panels, you'll help ensure their maximum efficiency and prolong their lifespan.

INVERTER AND CHARGE CONTROLLER MAINTENANCE: SAFEGUARDING YOUR POWER CONVERSION AND STORAGE

Inspecting and Maintaining Inverters

Your inverter plays a crucial role in converting the direct current (DC) generated by your solar panels into alternating current (AC) for use in your home or grid connection. Regular inspection and maintenance of your inverter are essential to ensure its optimal performance and longevity. Schedule routine inspections every six months or according to the manufacturer's recommendations, and be on the lookout for any warning signs, such as error codes, unusual noises, or overheating.

To maintain your inverter, follow these best practices:

- Keep the area around the inverter clean and free of dust, debris, and potential obstructions to airflow.
- Visually inspect the inverter for signs of damage, corrosion, or loose connections.
- Monitor the inverter's performance data, such as input and output power, efficiency, and any error codes.
- Consult the manufacturer's guidelines for specific maintenance tasks, such as firmware updates or internal component cleaning.

Charge Controller Care and Troubleshooting

Charge controllers are responsible for managing the flow of power between your solar panels, battery bank, and inverter. As with inverters, regular inspections and maintenance of your charge controller are crucial to ensure its proper functioning and extend its lifespan.

To care for your charge controller, follow these guidelines:

- Ensure that the charge controller is clean and free of dust or debris that could obstruct airflow or cause overheating.

- Check for loose connections or corrosion on terminals and wiring.
- Monitor the charge controller's performance data, such as charging current, battery voltage, and state of charge.
- Troubleshoot any issues by referring to the manufacturer's documentation and, if necessary, consult a professional for assistance.

Upgrading and Replacing Components as Needed

As your solar power system ages or your energy needs evolve, it may become necessary to upgrade or replace certain components, such as inverters and charge controllers. Regularly assessing the performance of these devices and comparing them to newer models on the market can help you determine when an upgrade or replacement is warranted. Upgrading to a more efficient or feature-rich device can improve your system's overall performance, potentially extending the life of your other components and enhancing your renewable energy experience.

By diligently maintaining your inverter and charge controller, you can ensure the continued efficiency and reliability of your solar power system, protecting your investment and maximizing the benefits of clean, renewable energy.

BATTERY MAINTENANCE: ENSURING LONGEVITY AND PERFORMANCE

Regular Battery Inspection and Cleaning

As an integral component of your solar power system, batteries require regular inspection and maintenance to ensure their longevity and optimal performance. It's important to establish a routine battery inspection schedule, ideally every three to six months, depending on the type and manufacturer's recommendations. During inspections, ensure that the battery terminals are clean and free of corrosion, which can negatively impact the battery's performance.

To clean battery terminals:

- Disconnect the battery from the system, following safety guidelines.
- Use a wire brush or a cloth soaked in a solution of water and baking soda to gently clean the terminals, removing any corrosion or dirt.
- Dry the terminals thoroughly and apply a thin layer of dielectric grease or petroleum jelly to protect them from future corrosion.
- Reconnect the battery, ensuring proper polarity and secure connections.

Monitoring Battery State of Charge and Capacity

Regularly monitoring the state of charge and capacity of your batteries is essential to avoid overcharging or deep discharging, which can shorten their lifespan.

Familiarize yourself with your battery's specifications, such as its nominal voltage, amp-hour capacity, and recommended depth of discharge.

Use a battery monitor or check the charge controller's display to track your battery's state of charge and ensure it remains within the recommended limits.

Additionally, perform periodic capacity tests to assess the battery's overall health and remaining life. If you notice a significant decline in capacity, it may be time to consider replacing the battery.

Understanding Battery Replacement and Recycling

Battery replacement is an inevitable part of maintaining a solar power system. Depending on the type and usage, batteries may need replacement every 5 to 15 years. A need for battery replacement may arise even sooner if the batteries are stored in harsh environmental conditions.

When it's time to replace your batteries, ensure that you choose new batteries that are compatible with your existing system, considering factors such as voltage, capacity, and chemistry.

Proper disposal of old batteries is crucial for both the environment and human health. Many batteries contain hazardous materials like lead, which can be harmful if not disposed of properly.

Research local recycling programs and regulations for battery disposal, and ensure that you recycle your old batteries in a responsible manner.

By regularly inspecting, cleaning, and monitoring your batteries, you can prolong their life and ensure the continued efficiency and reliability of your solar power system. Stay informed about battery replacement and recycling to minimize the environmental impact of your renewable energy system.

ELECTRICAL SYSTEM MAINTENANCE: KEEPING CONNECTIONS SAFE AND SECURE

Inspecting Wiring, Fuses, and Circuit Breakers

As an electrical engineer I've seen first-hand, the catastrophic consequences of failed electrical components. I can't stress enough the importance of regular electrical system maintenance to keep your solar power system safe, secure, and efficient. Start by inspecting your system's wiring, fuses, and circuit breakers every six months to one year, or more frequently if you observe any issues.

During inspections, look for any signs of wear, overheating or melting in addition to loose connections, or corrosion on wiring and connections. Check the tightness of terminal connections and ensure that they're secure. Examine fuses and circuit breakers for any visible damage or signs of tripping, which could indicate an underlying issue.

Identifying and Addressing Signs of Wear or Damage

Early identification and resolution of electrical system wear or damage can prevent more significant problems down the road. Some common signs of wear or damage include:

- Discolored or frayed wiring
- Loose or corroded connections
- Flickering or dimming lights
- Circuit breakers tripping frequently

If you encounter any of these issues, you may need to consult with a licensed electrician to diagnose the problem and make any necessary repairs or replacements. Addressing these issues promptly can prevent more severe damage and potential safety hazards.

Staying Up-to-Date with Electrical Safety Standards

Staying informed about the latest electrical safety standards and local regulations is essential to maintaining a safe and compliant solar power system. These standards may evolve over time, and it's crucial to keep your system up-to-date with any changes.

Regularly review the guidelines provided by your utility company, as well as any local, regional, or national electrical codes that apply to your system. Consult with a licensed electrician if you have questions or concerns about your system's compliance with current standards.

Maintaining your solar power system's electrical components is vital for ensuring safe and efficient operation. Regular inspections, identifying and addressing signs of wear or damage, and staying informed about electrical safety standards are all essential steps to keep your system functioning optimally and safely.

MONITORING SYSTEM PERFORMANCE: TRACKING YOUR SOLAR POWER SYSTEM'S HEALTH

Using Remote Monitoring Tools and Software

Closely monitoring your solar power system's performance is critical to ensure it remains efficient and effective. Remote monitoring tools and software can help you keep track of your system's health and identify potential issues early on. Many solar power systems come with built-in monitoring capabilities, or you can invest in third-party solutions.

These monitoring tools typically provide real-time data on system output, energy consumption, and other relevant metrics. You can often access this information through a web portal or mobile app, allowing you to keep a close eye on your system's performance from anywhere with internet access.

Analyzing Data for Trends and Anomalies

With access to detailed performance data, it's essential to analyze it for trends and anomalies. Regularly reviewing this information can help you spot potential issues before they become significant problems. Look for any sudden drops in output, unexpected fluctuations in energy consumption, or unusual patterns that could indicate an issue with your system.

By identifying trends and anomalies early, you can take appropriate action to address them, such as scheduling an inspection or maintenance visit, troubleshooting specific components, or making system adjustments as needed.

Making Informed Decisions Based on System Performance Data

Having a wealth of performance data at your fingertips can empower you to make informed decisions about your solar power system. Use this information to identify areas where your system could be improved or optimized, such as upgrading components, adjusting your system's configuration, or investing in energy storage solutions.

Monitoring your system's performance data can also help you determine when it's time to replace aging components, such as batteries or inverters, before they cause more significant issues. In addition, tracking your system's performance over time can provide valuable insights into its overall efficiency and return on investment.

In conclusion, monitoring your solar power system's performance is crucial to maintaining its health and efficiency. By using remote monitoring tools, analyzing data for trends and anomalies, and making informed decisions based on performance data, you can ensure your solar power system remains a valuable, reliable, and sustainable energy solution for years to come.

WORKING WITH PROFESSIONALS: WHEN TO SEEK EXPERT ASSISTANCE

Identifying Situations That Require Professional Help

While you can handle many aspects of solar power system maintenance yourself, certain situations require the expertise of a professional. These may include diagnosing complex system issues, performing major component replacements, or addressing problems that could pose a safety risk.

Some common situations that may require professional help include:

- Persistent or recurring system issues that cannot be resolved through routine maintenance
- Malfunctioning or damaged inverters, charge controllers, or other critical components
- Electrical problems that could pose a safety hazard, such as exposed wiring or damaged circuit breakers

Selecting and Working with Solar Power System Professionals

When it's time to call in professional help, it's crucial to choose a qualified and experienced solar power system professional. Look for technicians with appropriate certifications and a proven track record of success in the field. You can also seek recommendations from friends, family, or online reviews to help you find a reliable professional.

Once you've selected a solar power system professional, it's essential to communicate your concerns and provide them with relevant information about your system. This can help them accurately diagnose and resolve any issues you're facing.

Maintaining Warranty Coverage and Compliance

Many solar power system components come with warranties that protect you against defects or malfunctions. To maintain these warranties, it's essential to follow the manufacturer's guidelines for maintenance and repairs. In many cases, this means working with professionals for specific tasks, such as installing or replacing key components.

Additionally, some local regulations or utility company policies may require you to work with licensed professionals for certain tasks, such as grid interconnection or electrical work. Ensuring compliance with these requirements can help you avoid potential penalties or complications down the road.

Recognizing when to seek professional help is crucial for maintaining the safety and performance of your solar power system. By identifying situations that require expert assistance, selecting qualified professionals, and maintaining warranty coverage and compliance, you can ensure you keep your system in top shape, generating clean, sustainable electricity for years to come.

CONCLUSION: EMBRACING A PROACTIVE APPROACH TO SOLAR POWER SYSTEM MAINTENANCE AND MONITORING

The Benefits of a Well-Maintained Solar Power System

A well-maintained solar power system offers numerous benefits that extend beyond just energy production. Proper maintenance helps ensure your system runs efficiently, maximizing your return on investment and minimizing energy costs. A well-maintained system is also more reliable, reducing the likelihood of unexpected downtime or component failures.

Additionally, taking care of your solar power system can help prolong the lifespan of your components, delaying the need for costly replacements. By addressing issues early on, you can prevent more significant problems from developing, ultimately saving you time and money in the long run.

Developing a Sustainable Maintenance and Monitoring Routine

To fully reap the benefits of a well-maintained solar power system, it's essential to develop a sustainable maintenance and monitoring routine. This includes performing regular inspections and cleanings, keeping an eye on system performance, and addressing any issues as they arise.

Incorporate the use of remote monitoring tools and software to track your system's performance and identify trends or anomalies. This data can help inform your maintenance decisions and allow you to address potential issues before they become major problems.

Remember that some situations may require professional assistance, so it's essential to know when to call in an expert. By working with qualified solar power system professionals, you can ensure that your system remains safe, efficient, and compliant with all relevant regulations and warranty requirements.

In summary, embracing a proactive approach to solar power system maintenance and monitoring is key to ensuring your system remains an efficient, reliable, and long-lasting energy solution. By taking care of your system and addressing issues early on, you can maximize your return on investment and enjoy the many benefits that solar power has to offer.

[CHAPTER 8] — THE SOLAR POWER SAVINGS BONANZA: MAXIMIZING YOUR GREEN ENERGY GAINS

Welcome to the solar power savings bonanza! By now, you've successfully designed and installed your solar power system, and you're well on your way to reaping the rewards of your green energy investment.

However, there's still more you can do to make the most of your solar power system. The secret to truly maximizing your gains lies in finding the perfect balance between financial savings and environmental advantages. In this chapter, we'll show you the steps you can take to optimize your system's performance, reduce your energy bills, and contribute to a more sustainable, eco-friendly future.

We'll dive into the world of efficiency, smart management, future-proof energy, and net metering, exploring how each of these topics can help you make the most of your solar power system. So, let's uncover the secrets to unlocking the full potential of your solar energy gains!

EFFICIENCY SUPERSTARS: SELECTING TOP-NOTCH SOLAR COMPONENTS

Investing in High-Efficiency Solar Panels

When it comes to solar panel selection, efficiency is a crucial factor. High-efficiency solar panels convert sunlight into electricity more effectively, allowing you to generate more power with less space. While they may have a higher initial cost, these panels often result in greater long-term savings, reduced system size, and a faster return on investment.

To identify high-efficiency panels, look for those with a higher conversion rate, usually above 20%. Research the best and most highly rated brands and models available in the market and be prepared to invest in quality panels that will deliver optimal performance for years to come.

Picking the Ideal Inverter for Your System

An inverter's main job is to convert the direct current (DC) generated by your solar panels into alternating current (AC) that can be used by your home or fed back into the grid.

The efficiency of an inverter directly impacts the overall efficiency of your solar system. When selecting an inverter, consider factors such as efficiency rating, compatibility with your solar panels, and the type of inverter (string, micro, or power optimizer).

Top-notch inverters typically have efficiency ratings above 95%, ensuring minimal energy losses during the conversion process. Keep in mind that a well-chosen inverter will contribute to the overall performance and reliability of your solar power system.

The Crucial Role of Charge Controllers and Batteries in System Efficiency

For off-grid or grid-tied systems with battery backup, charge controllers and batteries play a critical role in maintaining system efficiency. Charge controllers regulate the flow of electricity between the solar panels and the batteries, preventing overcharging and extending battery life. Selecting a high-quality charge controller, such as a maximum power point tracking (MPPT) controller, can improve system efficiency by up to 30% compared to more basic pulse width modulation (PWM) controllers.

Battery efficiency is also essential for energy storage and backup. High-quality, deep-cycle batteries designed for solar applications can store energy more efficiently and provide a longer lifespan.

When selecting batteries, consider factors such as capacity, depth of discharge, and cycle life. Lithium-ion batteries are often recommended for their high efficiency, long life, and low maintenance requirements, although other types of batteries, such as lead-acid or flow batteries, may also be suitable depending on your system's specific needs. By carefully choosing high-performance charge controllers and batteries, you can maximize the efficiency and longevity of your solar power system.

SMART MANAGEMENT: HARNESSING THE POWER OF SOLAR WITH ENERGY EFFICIENCY MEASURES

Energy Audits: Uncovering Hidden Opportunities for Savings

An energy audit is a thorough assessment of your home or business's energy use, which helps you identify areas where you can save energy and reduce consumption.

By conducting an energy audit, you can pinpoint inefficiencies in your building's envelope, heating and cooling systems, lighting, and appliances. By addressing these inefficiencies, you can improve the overall performance of your solar power system, reduce energy waste, and lower your utility bills.

Energy audits can be performed by professionals or through a DIY approach, using online resources and tools to guide the process.

Energy Efficiency Upgrades: Boosting Performance and Reducing Consumption

Once you've identified areas for improvement through an energy audit, it's time to implement energy efficiency upgrades. These upgrades can range from simple, low-cost measures like sealing air leaks, adding insulation, and switching to LED lighting, to more substantial investments such as upgrading HVAC systems, installing energy-efficient windows, or replacing outdated appliances.

By making these improvements, you'll not only enhance the comfort and value of your home or business but also maximize the return on your solar power investment by reducing the overall energy demand.

The Negawatt Revolution: Saving Energy by Using It Wisely

The concept of the "negawatt" is based on the idea that the most sustainable and cost-effective energy is the energy that's never used. By prioritizing energy efficiency and smart energy management, you can essentially "generate" negawatts, which contribute to reducing your energy bills and reliance on fossil fuels.

The “Negawatt Revolution” encompasses various strategies, such as efficiency measures and demand-side management, smart grids, and the use of energy management systems to monitor and control energy consumption. By adopting these smart management practices, you'll optimize the performance of your solar power system and maximize your green energy gains.

FUTURE-PROOF ENERGY: PREPARING FOR THE EVOLUTION OF SOLAR TECHNOLOGY

Staying Informed About Emerging Solar Technologies

As solar technology continues to advance at a rapid pace, staying informed about the latest developments is crucial for maximizing the benefits of your solar power system.

Keep an eye on industry trends, attend webinars or conferences, and follow reputable sources for updates on new solar technologies and breakthroughs. By staying informed, you'll be better equipped to make informed decisions about upgrading your system or incorporating new technologies when the time is right.

Evaluating the Feasibility of Solar System Upgrades

As new solar technologies become available, it's essential to evaluate the feasibility of upgrading your system to benefit from these advancements.

Consider factors such as cost, expected performance improvements, and compatibility with your existing setup. Some upgrades might be as simple as swapping out an older inverter for a more efficient model, while others could involve a more significant investment, such as adding energy storage or upgrading your solar panels.

When evaluating potential upgrades, be sure to weigh the benefits against the costs to determine if they're a worthwhile investment for your specific situation.

Planning for a Sustainable Energy Future

Embracing a sustainable energy future involves more than just installing a solar power system; it also requires a long-term commitment to energy efficiency, smart management, and staying up-to-date with the latest technological advancements.

As you plan for the future, consider ways to further reduce your energy consumption, increase the use of renewable energy sources, and contribute to a more resilient, decarbonized grid.

By taking a proactive approach to energy management and being open to new technologies, you'll be well-prepared for the ongoing evolution of solar technology and ready to reap the benefits of a cleaner, greener energy future.

NET METERING: TURNING YOUR SOLAR POWER SYSTEM INTO A REVENUE GENERATOR

Understanding Net Metering and Its Benefits, Including Its Function as an Infinite Battery

While not all utility companies offer net metering, it's essential to understand its benefits, including its function as an infinite battery. Your energy generation and usage levels can fluctuate throughout the year, and net metering can help you store excess energy in the grid. In essence, the grid acts as a giant annual battery that provides energy that you stored away when you had a surplus.

With net metering, the excess energy your solar panels generate is sent to the grid, and you receive credit for that energy. During times when your energy production is lower than your usage, such as at night or during cloudy days, you can draw energy from the grid using those credits. This helps you save on your electricity bill and makes your solar investment more cost-effective.

So, while net metering may not be available in all areas, it's worth exploring whether it's an option for you. By treating the grid as an infinite battery, you can maximize the benefits of your solar setup and save money in the long run.

Navigating the Net Metering Application Process

To participate in a net metering program, you'll need to apply through your utility company. The process typically involves submitting an application with information about your solar power system, such as the size, location, and interconnection details.

You may also be required to provide proof of insurance and sign a net metering agreement. Once your application is approved, your utility company will install a bi-directional meter at your property to measure both the energy you consume and the excess energy you feed back into the grid.

Maximizing Financial Returns Through Net Metering

To maximize the financial benefits of net metering, it's essential to optimize your solar power system's performance and ensure that you're generating as much clean energy as possible.

Strategies to increase your system's efficiency include proper solar panel placement and tilt, investing in high-quality components, and regularly maintaining and your system and cleaning your panels. Additionally, by implementing energy efficiency measures and smart energy management practices, you can further reduce your energy consumption, thereby increasing the amount of surplus energy you can sell back to the grid. By combining net metering with a well-designed and maintained solar power system, you can turn your green energy investment into a powerful revenue generator.

CONCLUSION: EMBRACING A HOLISTIC APPROACH TO SOLAR POWER SAVINGS AND SUSTAINABILITY

In conclusion, achieving maximum solar power savings and sustainability requires a holistic approach that considers every aspect of the system. From choosing the right equipment to proper installation, maintenance, and monitoring, each step is crucial to optimize the performance of your solar setup.

Investing in high-quality solar panels and equipment will save you money in the long run by minimizing maintenance and repair costs while improving the efficiency of your system. Ensuring proper installation, including the correct placement and orientation of your panels and appropriate wiring, can significantly affect the overall performance of your system.

Regular maintenance, including cleaning your panels and checking connections, is essential to keeping your system operating at peak efficiency. Monitoring your energy consumption and production can help identify potential issues early on and make adjustments to improve performance.

Finally, it's essential to consider the overall sustainability of your solar system, including end-of-life disposal and recycling of equipment. Opting for eco-friendly equipment and practices can minimize the environmental impact of your solar setup and contribute to a more sustainable future.

By embracing a holistic approach to solar power savings and sustainability, you can maximize the benefits of your solar investment while minimizing its impact on the environment.

[CHAPTER 9] — SOLAR SLEUTHING: TROUBLESHOOTING COMMON SOLAR POWER SYSTEM ISSUES

INTRODUCTION: BECOMING A SOLAR DETECTIVE

As a solar power system owner, it's essential to stay vigilant and proactive in detecting and resolving issues that may arise during the system's lifetime.

Like a skilled detective, you must be prepared to identify and understand the signs of potential problems, allowing you to take swift action when necessary. This chapter will guide you through the process of troubleshooting common solar power system issues, empowering you with the knowledge and confidence to address common challenges effectively.

Timely detection and resolution of system deficiencies are crucial for preserving your system's longevity, maintaining optimal system performance and maximizing your solar power system's benefits. Delays in identifying and addressing problems can lead to reduced energy generation, increased wear and tear on components, and, in some cases, safety hazards.

By staying observant and responsive to changes in your system's performance, you can minimize downtime and maximize the lifetime and efficiency of your solar investment.

Effective troubleshooting requires a solid understanding of your solar power system's components and their functions. This chapter will provide an overview of common issues that may arise in various parts of your solar power system, including solar panels, inverters, batteries, charge controllers, wiring, and monitoring equipment.

With the right knowledge in hand, you can confidently diagnose and resolve many issues on your own or make informed decisions about when to seek professional assistance. By embracing the role of a solar detective, you can take control of your solar power system's health and maximize its performance for years to come.

ADDRESSING ISSUES WITH SOLAR PANELS

Detecting and Remedying Underperforming Solar Panels

Solar panel performance can fluctuate due to various factors, such as age, temperature, and environmental conditions. Regular monitoring of your solar panels' output can help you identify underperforming panels early on. Compare the output of individual panels to their manufacturer specifications or to other similar panels in your system. If a panel consistently underperforms, it may be time for a thorough inspection, cleaning, or professional assessment.

Identifying and Fixing Mechanical or Electrical Damage

Mechanical or electrical damage to your solar panels can significantly impact their performance. Keep an eye out for visible signs of damage, such as cracked glass, broken cells, or damaged wiring.

If you detect any damage, promptly address the issue to prevent further degradation. In some cases, repairs may be possible, while in other cases, it may be more cost-effective to replace the damaged panel. Always consult a professional if you are unsure about the extent of the damage or the best course of action.

Resolving Shading and Soiling Concerns

Shading and soiling can both negatively impact the performance of your solar panels. Keep your solar array clear of obstructions, such as trees, buildings, or other objects that may cast shadows on your panels. Trim overhanging branches or consider relocating panels if shading becomes a persistent issue.

Dirt, dust, and debris can also accumulate on your solar panels over time, reducing their efficiency. Regularly inspect and clean your panels to ensure they remain free of grime. Use a soft brush or cloth and water to clean the surface gently. Avoid using abrasive materials or cleaning agents that may damage the panels.

In most cases, cleaning your panels once or twice a year is sufficient to maintain optimal performance. However, if you live in a particularly dusty or dirty environment, more frequent cleaning may be necessary.

INVERTER INTRICACIES: TACKLING ISSUES WITH YOUR POWER CONVERTER

Recognizing Common Inverter Error Codes and Messages

Inverters are equipped with diagnostic tools and display error codes or messages to help you identify potential issues. Familiarize yourself with your inverter's manual to understand the meaning of these codes and the recommended steps for resolving them.

Common issues may include overheating, grid connectivity problems, or internal component failures. If you cannot resolve an issue on your own, reach out to the inverter manufacturer or a professional solar technician for assistance.

Troubleshooting Inverter Performance and Connectivity

A well-functioning inverter is crucial for an efficient solar power system. If you notice a drop in system performance, check the inverter's display or monitoring platform for any issues. Ensure that all connections to the solar panels, battery, and grid are secure and free of corrosion or damage. Examine the inverter for any signs of physical damage, overheating, or unusual noises. If your inverter is connected to the internet for remote monitoring, verify that the connection is stable.

Ensuring Proper Inverter Settings and Firmware Updates

It's essential to keep your inverter settings configured correctly to match your solar panel system, battery, and grid requirements. Regularly review and update these settings as needed to ensure optimal performance.

Additionally, inverter manufacturers frequently release firmware updates to improve the device's functionality and address known issues. Periodically check for and install these updates to keep your inverter functioning at its best.

If you encounter difficulties with inverter settings or firmware updates, consult your inverter's manual or contact the manufacturer for guidance.

BATTERY BANK BLUES: DIAGNOSING AND RESOLVING ENERGY STORAGE PROBLEMS

Assessing Battery Health and State of Charge

Regularly monitoring your battery bank's health and state of charge is essential for maintaining an efficient solar power system. Most battery systems have built-in monitoring capabilities that display the state of charge, voltage, and other essential parameters. Check these indicators and compare them to the manufacturer's specifications to ensure your batteries are performing as expected. If you notice any discrepancies, investigate further to identify potential issues.

Addressing Capacity Loss and Premature Aging

Over time, batteries may lose capacity due to normal wear and tear or improper maintenance. To minimize capacity loss and extend battery life, follow the manufacturer's recommendations for charging and discharging, maintain appropriate temperature conditions, and conduct regular maintenance checks.

If your batteries are experiencing capacity loss or premature aging, consider investing in higher quality batteries or upgrading to more advanced energy storage technologies.

Safely Handling Battery Leaks and Swollen Cells

Battery leaks and swollen cells are hazardous situations that require immediate attention. If you notice any signs of leakage, swelling, or deformation in your battery bank, take the following steps:

1. Turn off your solar power system and disconnect the batteries to prevent further damage.
2. Use proper personal protective equipment, such as gloves and goggles, when handling damaged batteries.
3. Consult your battery's manual or contact the manufacturer for guidance on safely disposing of the damaged batteries.
4. Inspect the remaining batteries for any signs of damage and replace them if necessary.
5. Implement proper maintenance practices and monitor your battery bank regularly to prevent future issues.

CHARGE CONTROLLER CONUNDRUMS: KEEPING YOUR ENERGY REGULATOR IN CHECK

Understanding Charge Controller Error Codes and Indicators

Charge controllers are responsible for regulating the energy flow between your solar panels, battery bank, and loads. To keep your system functioning efficiently, it's crucial to understand the error codes and indicators displayed by your charge controller. Consult your charge controller's manual for a detailed explanation of these codes and learn how to identify potential issues promptly.

Troubleshooting Charging and Discharging Issues

If your charge controller isn't effectively managing the charging and discharging of your battery bank, several issues might be at play. These can include incorrect voltage settings, damaged or loose wiring connections, or malfunctioning components.

To troubleshoot these issues, perform a visual inspection of the charge controller and wiring, check the voltage settings, and test the solar panels and battery bank to ensure they are functioning correctly.

If the issue persists, contact the manufacturer or a solar professional for assistance.

Ensuring Correct Charge Controller Settings and Compatibility

To maximize your solar power system's efficiency, your charge controller must be configured correctly and compatible with your solar panels and battery bank. Double-check the manufacturer's recommendations for voltage, current, and other settings. Ensure that your charge controller is compatible with the type and size of your battery bank and solar panels. If you've recently upgraded or expanded your system, make sure to adjust the charge controller settings accordingly and verify that it is capable of handling the new components.

WIRING WORRIES: DETECTING AND FIXING ELECTRICAL SYSTEM ANOMALIES

Identifying Signs of Faulty or Damaged Wiring

Faulty or damaged wiring can significantly impact your solar power system's performance and pose safety risks. Signs of wiring issues include flickering lights, intermittent power loss, sparking, or discoloration around outlets and switches.

To identify and resolve such problems, perform regular visual inspections of your system's wiring, looking for signs of wear, damage, or corrosion. If you suspect an issue, consult a licensed electrician to assess and repair the problem.

Inspecting and Replacing Fuses and Circuit Breakers

Fuses and circuit breakers are essential safety components that protect your solar power system from overcurrent events. To ensure their proper functioning, inspect them regularly for signs of wear, damage, or tripping. If a fuse has blown or a circuit breaker has tripped, identify the cause of the overcurrent event and rectify it before replacing the fuse or resetting the breaker.

Consult your system's documentation for guidance on the appropriate fuse and circuit breaker ratings and replacement procedures.

Confirming Proper Grounding and Surge Protection

A well-grounded solar power system and adequate surge protection are critical for ensuring system safety and preventing damage from voltage spikes or lightning strikes. Regularly inspect your system's grounding connections, ensuring they are clean, tight, and corrosion-free. Check your surge protection devices for signs of wear or damage and replace them as needed. If you're unsure about the adequacy of your system's grounding or surge protection, consult a solar professional or licensed electrician for guidance.

MONITORING MISSTEPS: ENSURING ACCURATE SYSTEM PERFORMANCE DATA

Addressing Communication Issues Between System Components

Effective communication between your solar power system's components is essential for accurate monitoring and optimal performance. To address communication issues, regularly inspect wiring connections between devices, ensuring they are secure and free from damage. Check for software or firmware updates for your inverter, charge controller, and monitoring devices, as outdated software can sometimes cause communication problems. If issues persist, consult your system's documentation or contact the manufacturer for assistance.

Calibrating and Verifying Monitoring Sensors and Devices

Proper calibration and verification of monitoring sensors and devices are crucial for obtaining accurate performance data. Periodically check the accuracy of sensors, such as temperature, irradiance, and current sensors, by comparing their readings with known values or reference devices.

Consult your system's documentation for guidance on calibration procedures and recommended calibration intervals. Replace or recalibrate any sensors or devices that are found to be inaccurate.

Interpreting and Acting Upon System Performance Data

System performance data can provide valuable insights into the health and efficiency of your solar power system, but only if you know how to interpret it correctly. Familiarize yourself with key performance indicators, such as energy production, system efficiency, and state of charge, and compare them to expected or benchmark values. Analyze trends and anomalies in the data to identify potential issues or areas for improvement. If you're unsure how to interpret your system's data or need assistance identifying appropriate actions to optimize performance, consider consulting a solar professional.

CONCLUSION: MASTERING SOLAR SYSTEM TROUBLESHOOTING AND MAINTENANCE

As a solar power system owner, adopting a proactive approach to system care and troubleshooting is essential to ensure optimal performance and longevity. Regularly inspecting and maintaining your system components will help prevent potential issues and ensure that your investment continues to provide clean, green energy for years to come.

Understanding the basics of solar system troubleshooting empowers you to address minor issues and concerns as they arise, while also helping you recognize when professional assistance is necessary for more complex problems. By staying informed about common solar power system issues, you can confidently tackle challenges and make informed decisions about your system's maintenance and performance.

Ultimately, mastering solar system troubleshooting and maintenance not only safeguards your investment but also contributes to a sustainable energy future.

[CHAPTER 10] — THINK BIGGER: EXPANDING YOUR SOLAR POWER SYSTEM FOR A GREENER TOMORROW

INTRODUCTION: EMBRACING SOLAR EXPANSION FOR A BRIGHTER TOMORROW

Once you've settled into your solar power system it's common for the question to arise, "How can I make this even more badass?" In this chapter, we will explore the opportunities for expanding your solar power system once you're ready to add more green octane.

Assessing your energy needs and solar potential is the first step in determining whether expanding your solar power system is a viable option. By carefully examining your energy consumption patterns, you can identify areas where increased solar generation capacity could make a difference in your overall energy usage. A thorough evaluation of your existing solar system and the available space for expansion will provide valuable insights into the feasibility of system expansion.

In this chapter, we will guide you through the process of evaluating, planning, and implementing a solar power system expansion, ensuring that you can make the most of your renewable energy investments. We will also touch upon additional renewable energy options that can help you diversify your portfolio.

EVALUATING YOUR CURRENT SYSTEM: ASSESSING EXPANSION OPPORTUNITIES

Reviewing Your Energy Consumption Patterns

Understanding your energy consumption patterns is the first step in assessing expansion opportunities for your solar power system. Analyze your monthly energy bills and usage data to identify any changes in consumption over time. This information can help you determine the ideal size and capacity for an expanded solar power system.

Identifying System Performance Trends and Areas for Improvement

Regularly monitoring your solar power system's performance will provide valuable insights into its efficiency and effectiveness. Analyze trends in power production and identify any inconsistencies or inefficiencies that may be limiting your system's potential. Investigating these issues can help you target areas for improvement, such as upgrading components or repositioning solar panels for better sun exposure.

Analyzing the Scalability of Your Existing Solar Power System

Before you can expand your solar power system, you must first assess its scalability. Review the specifications of your current components, including solar panels, inverters, and charge controllers, to determine if they can support additional capacity.

Also, consider the available space for additional panels and the structural integrity of your mounting system. Understanding the limitations and potential of your existing system will guide your expansion planning and ensure a smooth transition to a larger, more efficient solar power setup.

SYSTEM EXPANSION STRATEGIES: FINDING THE RIGHT APPROACH

When looking to expand your solar power system, there are several strategies to consider. This section will discuss three key approaches, including increasing generation capacity through additional solar panels, expanding battery storage for greater energy independence, and integrating smart energy management solutions for optimal efficiency.

Adding More Solar Panels to Increase Generation Capacity

One straightforward way to expand your solar power system is by adding more solar panels. By increasing the number of panels, you can harness more sunlight and generate additional electricity to meet your energy needs.

When considering this approach, consider the available space for new panels and the compatibility of your existing components. Also, ensure that your mounting system can accommodate the added weight and structural requirements.

Expanding Battery Storage for Greater Energy Independence

Expanding your battery storage capacity allows you to store more solar-generated electricity, reducing your reliance on grid power and enhancing your energy independence. This approach is particularly beneficial for those with high energy demands or in areas with frequent power outages.

Assess the capacity and performance of your current battery bank and consider upgrading to higher-capacity or more advanced battery technologies. Remember to ensure compatibility with your existing system components.

Integrating Smart Energy Management Solutions

Incorporating smart energy management solutions into your solar power system can optimize energy usage and further increase your overall system efficiency.

These solutions may include smart thermostats, energy-efficient appliances, and home automation systems that intelligently control lighting, heating, and cooling. Implementing these technologies can help you maximize the benefits of your solar power system by reducing energy waste and ensuring that the generated electricity is utilized in the most efficient manner possible.

PLANNING FOR A SUCCESSFUL EXPANSION: KEY CONSIDERATIONS

When planning to expand your solar power system, it is essential to consider several factors that will ensure a successful and seamless process. This section will discuss three critical considerations: accurately sizing your expanded solar power system, navigating permits and regulations, and budgeting and financing your system's growth.

Sizing Your Expanded Solar Power System Accurately

To optimize your expanded solar power system, it is crucial to accurately size it based on your current and future energy needs. Analyze your energy consumption patterns and consider any changes in usage that may occur, such as adding new appliances or an electric vehicle.

Additionally, consider any energy efficiency measures you have implemented or plan to introduce, as they can impact your overall energy demand. Consult with a solar professional to help determine the appropriate size for your expanded system, considering factors like available space, mounting options, and component compatibility.

Navigating Permits and Regulations for System Expansion

As you plan your solar power system expansion, it's essential to be aware of any permits, regulations, and codes that may apply. These requirements can vary by location and may include building permits, electrical permits, and interconnection agreements with your utility. Review your local and state regulations, consult with a solar professional, and ensure that your expanded system complies with all necessary codes and standards to avoid any complications or delays in the expansion process.

Budgeting and Financing Your Solar Power System Growth

Expanding your solar power system requires a financial investment, and it's crucial to budget and plan for these costs. Review your current system's costs and obtain quotes for the additional components and installation services needed for expansion. Consider various financing options, such as solar loans, power purchase agreements (PPAs), or leasing programs, to help you fund your system's growth.

Also, research available incentives, rebates, and tax credits that can help offset expansion costs and make your investment even more worthwhile. By carefully planning and budgeting for your solar power system expansion, you can ensure a smooth process and maximize the return on your investment.

IMPLEMENTING YOUR EXPANSION PLAN: FROM DESIGN TO INSTALLATION

Expanding your solar power system is an exciting endeavor that can lead to increased energy savings and a reduced carbon footprint. However, to ensure a successful expansion, it's crucial to carefully implement your plan from design to installation. In this section, we'll discuss collaborating with solar professionals for seamless integration, ensuring compatibility and efficiency of new components, and preparing for installation while minimizing system downtime.

Collaborating with Solar Professionals for Seamless Integration

Working closely with solar professionals is essential for a successful solar power system expansion. Solar experts can provide valuable insights into system design, component selection, and installation best practices. They will help you navigate the expansion process, ensuring your new components are seamlessly integrated into your existing system, and any potential issues are addressed before they become problems. Choose a reputable solar professional with experience in system expansion projects to guarantee a smooth and successful implementation.

Ensuring Compatibility and Efficiency of New Components

When expanding your solar power system, it's vital to ensure that the new components are compatible with your existing setup. This includes matching the voltage and power ratings of new solar panels, inverters, and batteries to your current system.

Moreover, consider investing in high-efficiency components that will maximize your system's energy production and contribute to its overall performance. Consult with solar professionals to select the best components for your expansion project and verify their compatibility and efficiency.

Preparing for Installation and Minimizing System Downtime

The installation phase of your solar power system expansion should be planned carefully to minimize any disruption to your current system's operation. Coordinate with your solar professionals to schedule the installation during a time when your energy usage is low, such as weekends or holidays. Additionally, ensure all necessary permits, materials, and equipment are in place before the installation process begins.

During the installation, closely monitor your solar power system to minimize downtime and address any issues promptly. Communicate with the installation team regularly, and make sure they are aware of any site-specific constraints or requirements.

By thoroughly preparing for the installation and working closely with solar professionals, you can ensure a smooth expansion process and quickly start enjoying the benefits of your upgraded solar power system.

BEYOND SOLAR EXPANSION: EXPLORING ADDITIONAL RENEWABLE ENERGY OPTIONS

While expanding your solar power system can significantly increase your renewable energy production, there are other sustainable options worth considering. By diversifying your renewable energy portfolio, you can enhance your overall energy independence and reduce your carbon footprint even further. In this section, we will explore the potential of harnessing wind power, investigating small-scale hydroelectric solutions, and integrating electric vehicles with your solar power system.

Harnessing Wind Power for a Complementary Energy Source

Wind power can serve as an excellent complementary energy source to solar. Wind turbines capture the kinetic energy of the wind and convert it into electricity, often generating power during times when solar production is low, such as cloudy days or at night.

Small-scale wind turbines can be installed on residential properties, providing an additional source of clean energy. Consult with a renewable energy expert to determine the feasibility of adding wind power to your property and its potential benefits.

One drawback of DIY wind power, in contrast to solar, is that it relies on moving parts to generate power. This makes the system more susceptible to wear and degradation over time.

Investigating the Potential of Small-Scale Hydroelectric Solutions

For property owners with access to flowing water, small-scale hydroelectric systems can offer an additional source of renewable energy. Micro-hydro systems use the kinetic energy of moving water to generate electricity, producing power around the clock.

The viability of a small-scale hydroelectric system depends on factors such as water flow, head height, and local regulations.

Consult with a hydroelectric professional to assess the potential of incorporating this technology into your renewable energy mix.

Integrating Electric Vehicles and Solar Power

Electric vehicles (EVs) are threatening to replace internal combustion engines sooner than later. By integrating EV charging with your solar power system, you can utilize your solar energy production to power your vehicle, further reducing your dependence on fossil fuels.

This synergy can be optimized by installing an EV charger with smart features, allowing you to charge your vehicle during periods of high solar production or when electricity rates are low. Evaluate the benefits of incorporating an EV into your lifestyle and consult with solar professionals to ensure a seamless integration with your solar power system.

CONCLUSION: SECURING YOUR ENERGY FUTURE THROUGH SYSTEM EXPANSION

As we conclude this chapter on expanding your solar power system, it's essential to recognize the numerous benefits you can gain from system flexibility, accommodating future needs, and diversifying your generation sources. By staying proactive and planning for expansion, you'll be better prepared to achieve energy security for you and your family.

Increasing the capacity of your solar power system ensures that you can meet your growing energy demands while maintaining a reliable source of clean energy. This not only enables you to maintain your energy independence but also enhances your resilience in the face of fluctuating energy prices or grid outages.

System flexibility allows you to adapt to changing circumstances, such as adding new appliances or expanding your living space. By designing your solar power system with the potential for expansion in mind, you can easily accommodate future energy needs and avoid the costs and complications of a complete system overhaul.

Diversifying your generation sources by incorporating complementary technologies like wind or hydro power further bolsters your energy security. By tapping into multiple renewable energy sources, you'll be able to optimize your power generation and reduce your reliance on any single source, ensuring a more consistent energy supply.

In summary, expanding your solar power system is a strategic investment in your long-term energy security and self-sufficiency. By embracing system flexibility, planning for future needs, and diversifying your energy sources, you'll be well-equipped to face the challenges of an ever-evolving energy landscape while continuing to lead a sustainable lifestyle.

[CHAPTER 11] - CASHING IN ON DIY SOLAR: FINANCIAL INCENTIVES, REBATES, AND TAX CREDITS FOR SELF-INSTALLED SOLAR SYSTEMS

CAPITALIZING ON THE FINANCIAL PERKS OF DIY SOLAR

As a DIY solar system owner, you have already taken the initiative to embrace renewable energy and embark on a journey towards self-sustainability. In doing so, you've not only contributed to a greener environment but also unlocked several cost-saving advantages that come with self-installed solar systems.

This chapter aims to help you navigate the world of financial incentives, rebates, and tax credits specifically designed for DIY solar projects, ensuring you reap the maximum rewards from your investment.

The Cost-Saving Advantages of Self-Installed Solar Systems

By taking on the challenge of installing your solar system, you've significantly reduced the overall cost of your project. By eliminating or minimizing labor expenses and potentially sourcing components at more competitive prices, you're well-positioned to achieve a quicker return on your investment. The financial benefits don't end there, as there are numerous incentives, rebates, and tax credits that cater specifically to DIY solar projects.

Navigating Incentives, Rebates, and Tax Credits Tailored for DIY Solar

Financial support for solar energy projects comes in various forms, ranging from federal tax credits to state-level incentives and utility company rebates. For DIY solar system owners, understanding and capitalizing on these financial incentives can lead to substantial savings and accelerate the payback period of your investment.

The Inflation Reduction Act

The Inflation Reduction Act has made significant changes to residential solar incentives, providing increased savings for homeowners who install solar panels and/or energy storage.

The act extends the federal solar tax credit, making it more accessible to a larger number of homeowners. Additionally, the act provides more funding for state-level rebates, tax credits, and grants that are designed to incentivize DIY solar installations. Combining these incentives with local utility programs can significantly boost savings for DIY solar enthusiasts.

In the sections that follow, we will discuss the specifics of these incentives, eligibility requirements, and the process of claiming them, helping you unlock the full financial potential of your DIY solar power system.

STATE AND LOCAL SOLAR INCENTIVES: BOOSTING SAVINGS FOR DIY SOLAR ENTHUSIASTS

When considering solar power as a DIY project, it is essential to be aware of the various state and local incentives that can significantly boost your savings. These incentives can come in the form of rebates, tax credits, grants, and other programs, which can help you offset the initial costs of your solar installation.

Discovering State-Level Rebates, Tax Credits, and Grants Designed for DIY Solar

Many states offer rebates, tax credits, and grants to homeowners who install solar systems. These incentives vary from state to state, and some states offer more incentives than others. Therefore, it is crucial to research the specific incentives available in your state.

For instance, some states offer a percentage-based tax credit on the total cost of the solar installation. Others offer a flat-rate rebate, while some provide grants that cover a portion of the costs. Some states even offer cash incentives for homeowners who generate excess solar power and sell it back to the grid.

Identifying Local Utility Incentives and Programs for Self-Installed Systems

In addition to state-level incentives, local utilities may also offer incentives and programs for self-installed solar systems. These incentives can range from net metering programs to buy-back programs for excess power generated by your solar system.

As we described earlier in the book, net metering programs allow homeowners to receive credits on their utility bills for the excess solar power they generate and feed back into the grid. In contrast, buy-back programs allow homeowners to sell their excess solar power to their utility provider. Some utilities may also offer special rates for homeowners who generate their own solar power.

To maximize your payback, it is crucial for DIY solar enthusiasts to research and take advantage of state and local incentives to maximize their savings. By discovering state-level rebates, tax credits, and grants designed for DIY solar and identifying local utility incentives and programs for self-installed systems, you can make the most of your solar investment.

SOLAR RENEWABLE ENERGY CERTIFICATES (SRECS) FOR DIY SOLAR: MONETIZING YOUR GREEN ENERGY PRODUCTION

Understanding SRECs And Their Role In Renewable Energy Markets for DIY Solar

Solar Renewable Energy Certificates (SRECs) are a valuable financial incentive that allows solar system owners, including those with DIY installations, to monetize their green energy production. SRECs are tradable commodities that represent the environmental attributes of one megawatt-hour (MWh) of solar electricity generated. Utilities in certain states are required to meet Renewable Portfolio Standard (RPS) targets by purchasing SRECs from solar system owners.

As a DIY solar enthusiast, you can benefit from SREC markets by generating and selling SRECs associated with your solar system's energy production. The revenue from SREC sales can help offset the cost of your system and contribute to faster payback periods.

Selling SRECs to Generate Additional Income from Your Self-Installed Solar System

To participate in the SREC market, start by registering your DIY solar system with your state's relevant agency or authority. Once registered, you will be eligible to generate SRECs based on your system's production. Keep in mind that SREC markets exist only in certain states, and their values can fluctuate depending on supply and demand.

There are several ways to sell your SRECs:

Brokerage services: Some companies specialize in trading SRECs and can handle the entire process for you, from registration to sales. They may charge a fee or take a percentage of the revenue from SREC sales.

Direct sales: You can sell SRECs directly to utilities or other buyers through online platforms or auctions. This option may require more effort but can potentially result in higher returns.

Long-term contracts: Some utilities offer long-term contracts for SREC purchases, which can provide a steady and predictable income stream. These contracts typically last several years and involve selling SRECs at a fixed price.

By participating in the SREC market, you can further enhance the financial benefits of your DIY solar system. Selling SRECs allows you to generate additional income from your green energy production, helping to make your solar investment even more rewarding and environmentally impactful.

CONCLUSION: SEIZING THE FINANCIAL REWARDS OF DIY SOLAR

Going solar is not only an environmentally responsible choice, but it can also be a financially rewarding one. By taking advantage of various financial incentives, rebates, and tax credits, you can significantly reduce the cost of your solar power system and increase the return on your investment.

In this chapter, we have explored the different financial incentives available for DIY solar projects, from the federal Inflation Reduction Act (IRA) to state and local incentives and Solar Renewable Energy Certificates (SRECs). Each of these incentives has the potential to contribute to your overall savings and help you realize the full potential of your self-installed solar system.

As you navigate the world of DIY solar, make sure to stay informed about the available financial incentives and how to access them. Keep in mind that programs and incentives may change over time, so it is crucial to stay updated on the latest developments. By maximizing these financial perks, you can not only reduce the cost of your solar project but also become an advocate for solar power and sustainable living.

Embrace the financial benefits that come with DIY solar and continue to be an example of how renewable energy can play a significant role in creating a greener tomorrow.

[GLOSSARY] – SOLAR ENERGY TERMS: A HANDY REFERENCE FOR THE SOLAR ENTHUSIAST

AC (Alternating Current)

Alternating Current is an electrical current that periodically reverses its direction, typically following a sine wave pattern. In residential and commercial applications, AC is the most common form of electricity, as it can be easily transformed and transmitted over long distances with minimal loss.

Ampere (Amp)

Ampere, commonly referred to as an amp, is the unit of measurement for electric current. It represents the flow of electrons in a conductor, such as a wire, and is used to quantify the rate at which electrical charge passes through a point in a circuit.

Angle of Incidence

The angle of incidence refers to the angle between the incoming sunlight and the surface normal (the line perpendicular to the surface) of a solar panel or any other surface. This angle impacts the amount of sunlight directly hitting the solar panel, which affects its energy output.

Array

A solar array is a collection of multiple solar panels or modules electrically connected together to form a larger energy-generating unit. Arrays are designed to increase the overall power output and can be tailored to match specific energy requirements and available space.

Azimuth Angle

The azimuth angle is the horizontal angle measured between the sun's position in the sky and a reference direction, usually true north. In the context of solar energy, azimuth angle is essential for determining the optimal orientation of solar panels to maximize their exposure to sunlight throughout the day.

Balance of System (BOS)

Balance of System typically refers to all the components of a solar power system, excluding the solar panels themselves. These components can include inverters, charge controllers, batteries, wiring, mounting hardware, and other necessary electrical and mechanical equipment. BOS components play a critical role in the overall efficiency, safety, and performance of a solar power system.

Battery Bank

A battery bank is a group of interconnected batteries that store electrical energy produced by a solar power system. Battery banks allow for the storage of excess solar-generated electricity, which can be used when sunlight is not available, such as during nighttime or periods of low solar irradiance.

Battery Capacity

Battery capacity is a measure of the total amount of electrical energy that a battery can store, typically expressed in ampere-hours (Ah). A higher battery capacity indicates a greater ability to store energy and provide power over an extended period.

Bifacial Solar Panels

Bifacial solar panels are designed to capture sunlight from both their front and rear surfaces. These panels feature solar cells on both sides, allowing them to produce electricity from direct sunlight, as well as reflected and diffused light from the surrounding environment. Bifacial solar panels can potentially generate more energy than traditional monofacial panels.

Building-Integrated Photovoltaics (BIPV)

Building-Integrated Photovoltaics refers to the incorporation of solar panels directly into the building's structure, such as rooftops, facades, or windows. BIPV systems not only generate electricity but also serve as an integral part of the building's design and architecture, offering a visually appealing and space-saving alternative to traditional solar panel installations.

Charge Controller

A charge controller is a critical component in a solar power system, responsible for regulating the flow of electrical current between the solar panels and the battery bank. It prevents overcharging and undercharging of the batteries by controlling the charging and discharging process, ultimately extending the battery life and improving overall system efficiency.

Circuit Breaker

A circuit breaker is a safety device used in electrical systems to protect against electrical overloads and short circuits. When the current flowing through a circuit exceeds a predetermined limit, the circuit breaker trips and interrupts the flow of electricity, preventing damage to equipment and reducing the risk of fire.

Crystalline Silicon (c-Si)

Crystalline silicon is a widely used semiconductor material in the production of solar cells. It comes in two main forms: monocrystalline silicon and polycrystalline silicon. Crystalline silicon solar cells are known for their high efficiency, durability, and long lifespan, making them a popular choice for solar power installations.

Current

Electrical current is the flow of electric charge in a conductor, such as a wire. It is measured in amperes (amps) and represents the movement of electrons within a circuit. Current can be direct (DC), where the flow of electrons is in one direction, or alternating (AC), where the flow of electrons periodically reverses direction.

DC (Direct Current)

DC (Direct Current): Direct Current is the unidirectional flow of electric charge in a circuit. In a DC circuit, electrons consistently move in one direction, maintaining a constant voltage polarity. Solar panels produce DC electricity, which is then converted into AC electricity by an inverter to be used by household appliances and connected to the electric grid.

DC Disconnect

A DC Disconnect is a safety device in a solar power system that allows for the isolation of the DC side of the system, such as solar panels and batteries, from the inverter and other AC components. This enables safe maintenance, inspection, or troubleshooting of the system by cutting off the flow of electricity and reducing the risk of electrical shock.

Deep-Cycle Battery

Deep-cycle batteries are designed to provide a consistent flow of electricity over an extended period and to withstand regular deep discharges without significant damage. These batteries are commonly used in solar power systems, electric vehicles, and other applications that require a stable and reliable energy source over time.

Degradation Rate

The degradation rate refers to the gradual decline in the power output of a solar panel over its operational lifespan. Solar panels typically experience a small reduction in efficiency each year due to various factors, such as aging, weathering, and exposure to sunlight. The degradation rate is usually expressed as a percentage per year.

Distributed Generation

Distributed generation, also known as decentralized generation, refers to the production of electricity from small-scale, local energy sources that are connected directly to the consumer or the distribution network. Examples of distributed generation systems include rooftop solar panels, wind turbines, and small-scale hydroelectric installations. Distributed generation can help reduce transmission losses, increase energy efficiency, and enhance the reliability and resiliency of the electric grid.

Efficiency

In the context of solar power systems, efficiency refers to the ability of solar panels or other system components to convert sunlight into usable electrical energy. Solar panel efficiency is typically expressed as a percentage and represents the proportion of the solar energy incident on the panel that is converted into electricity.

Electric Grid

The electric grid, or power grid, is a complex network of power plants, substations, transformers, and transmission and distribution lines that deliver electricity from power generation facilities to homes, businesses, and other consumers. The electric grid ensures a continuous and reliable supply of electricity to meet the varying demands of consumers.

Energy Audit

An energy audit is a systematic evaluation of a building's energy use, aiming to identify opportunities for energy savings and efficiency improvements. Energy audits typically involve an assessment of the building's energy consumption patterns, thermal performance, and HVAC systems, as well as an examination of potential upgrades and their associated costs and benefits.

Energy Conservation

Energy conservation refers to the practice of reducing energy consumption by using resources more efficiently, minimizing waste, and adopting energy-saving measures. This can be achieved through behavioral changes, implementing energy-efficient technologies, and optimizing the design and operation of buildings and appliances. Energy conservation not only helps to lower energy costs but also contributes to environmental protection by reducing greenhouse gas emissions.

Energy Independence

Energy independence is the ability of a household, business, or country to meet its energy needs without relying on external sources or imports. This can be achieved by utilizing domestic renewable energy resources, such as solar, wind, or hydroelectric power, increasing energy efficiency, and reducing overall energy consumption. Energy independence can provide economic benefits, enhance energy security, and contribute to reducing greenhouse gas emissions.

Energy Storage

Energy storage refers to the capture and retention of energy produced at one time for use at a later time. In the context of solar power systems, energy storage typically involves the use of batteries to store excess electricity generated by solar panels during the day, which can then be used during periods of low sunlight or at night. Energy storage technologies can help improve the reliability, efficiency, and flexibility of renewable energy systems.

Feed-in Tariff (FIT)

A feed-in tariff is a policy mechanism designed to promote the adoption of renewable energy sources by guaranteeing a fixed price for the electricity generated by these sources and fed into the electric grid. FITs typically involve long-term contracts and can provide a stable revenue stream for renewable energy producers, such as solar power system owners, encouraging investment in clean energy technologies.

Fuse

A fuse is an electrical safety device that protects a circuit from excessive current by melting a metal filament or wire when the current flow exceeds a predetermined threshold. When the fuse "blows" or "trips," it interrupts the flow of electricity and prevents potential damage to electrical components or risk of fire. Fuses are used in solar power systems to safeguard components like solar panels, inverters, and batteries.

Grid-Connected System

A grid-connected solar power system, also known as a grid-tied system, is a solar installation that is connected to the electric grid, allowing the system owner to draw electricity from the grid when solar generation is insufficient to meet demand and to feed excess solar-generated electricity back into the grid. Grid-connected systems can help reduce electricity bills and provide a stable power supply while contributing to a cleaner energy mix.

Grounding

Grounding, or earthing, is the process of creating a direct, low-resistance connection between an electrical system and the earth. This connection provides a safe pathway for fault currents to dissipate, reducing the risk of electrical shock, fire, or equipment damage. In solar power systems, grounding is essential for safety and proper system operation, ensuring that voltage fluctuations and transient events, such as lightning strikes, do not harm the system components.

Hybrid Solar System

A hybrid solar system combines the features of grid-tied and off-grid solar systems. It includes solar panels, an inverter, a battery bank, and a connection to the electrical grid. This configuration allows the system to draw power from the grid when necessary, store excess solar-generated electricity in batteries for later use, and supply power to the grid when generation exceeds consumption. Hybrid systems can offer greater energy independence and reliability compared to grid-tied systems.

Hydroelectric Power

Hydroelectric power is a form of renewable energy that generates electricity by harnessing the kinetic energy of flowing or falling water. This energy is typically captured by constructing dams or other structures that control water flow and drive turbines connected to electrical generators.

Inverter

An inverter is an electrical device that converts direct current (DC) electricity produced by solar panels into alternating current (AC) electricity, which is the standard form used by household appliances and the electric grid. Inverters are essential components of solar power systems, enabling the harvested solar energy to be used efficiently and safely.

Irradiance

Irradiance is a measure of the power of solar radiation per unit area, typically expressed in watts per square meter (W/m^2). It represents the amount of sunlight available for solar energy production at a given location and time. Irradiance is influenced by factors such as geographic location, time of day, season, and atmospheric conditions, and is a critical parameter for determining the performance and efficiency of solar power systems.

Kilowatt (kW)

A kilowatt (kW) is a unit of power equivalent to one thousand watts (W). Power represents the rate at which energy is produced or consumed. In the context of solar power systems, kilowatts are commonly used to express the capacity or output of solar panels, inverters, and other system components.

Kilowatt-hour (kWh)

A kilowatt-hour (kWh) is a unit of energy equivalent to one kilowatt (1 kW) of power consumed or produced for one hour. In the context of solar power systems and electricity usage, kilowatt-hours are used to measure energy production or consumption over a period of time. This unit is commonly used for billing purposes and to quantify the performance and savings of solar power systems.

Lead-Acid Battery

A lead-acid battery is a type of rechargeable battery commonly used in solar power systems for energy storage. It consists of lead plates and an electrolyte solution of sulfuric acid, which undergoes chemical reactions to store and release electrical energy. Lead-acid batteries are known for their affordability, reliability, and ease of recycling. However, they typically have lower energy density, shorter cycle life, and require more maintenance compared to lithium-ion batteries.

Lithium-Ion Battery

A lithium-ion battery is a type of rechargeable battery widely used in solar power systems for energy storage, as well as in electric vehicles and portable electronic devices. It relies on the movement of lithium ions between a positive and a negative electrode, with an electrolyte facilitating the ion transfer. Lithium-ion batteries offer high energy density, long cycle life, and low maintenance, but they can be more expensive and require specific safety precautions compared to lead-acid batteries.

Load

In the context of electrical systems, a load refers to any device or appliance that consumes electrical energy, such as lights, refrigerators, or air conditioners. In solar power systems, the load is an essential factor when sizing and designing the system, as it directly impacts the amount of solar generation capacity and energy storage required to meet the user's energy needs.

Maximum Power Point Tracking (MPPT)

Maximum Power Point Tracking (MPPT) is a technology used in solar charge controllers to optimize the power output of solar panels. MPPT algorithms continuously adjust the electrical load connected to the panels, ensuring that they operate at their maximum power point under varying sunlight and temperature conditions. This results in higher energy production and increased overall system efficiency compared to traditional pulse width modulated (PWM) charge controllers.

Microinverter

A microinverter is a small inverter designed to convert the direct current (DC) electricity produced by a single solar panel into alternating current (AC) electricity. In a solar power system with microinverters, each panel has its own dedicated inverter, allowing for independent operation and improved system performance. Microinverters can help mitigate the impact of shading, soiling, or panel mismatch, and can simplify system design and installation compared to traditional string inverters.

Monocrystalline Silicon (Mono-Si)

Monocrystalline silicon, or mono-Si, is a type of crystalline silicon used to manufacture solar cells and panels. Mono-Si cells are created from a single, continuous crystal structure, resulting in a uniform appearance and higher efficiency compared to other silicon types, such as polycrystalline silicon. Monocrystalline solar panels typically have a longer lifespan, better performance in low light, and a smaller footprint due to their higher efficiency, but they can be more expensive than other panel types.

Mounting System

A mounting system is a structural framework used to secure solar panels to a surface, such as a roof, ground, or pole. Mounting systems are essential components of solar power installations, providing stability, proper panel orientation, and ventilation for optimal performance and longevity. Various mounting system designs are available to accommodate different roof types, site conditions, and aesthetic preferences, including fixed, adjustable, and tracking systems.

National Electrical Code (NEC)

The National Electrical Code (NEC) is a set of safety standards and guidelines for the installation and maintenance of electrical systems in the United States. Developed and published by the National Fire Protection Association (NFPA), the NEC provides best practices for electrical wiring, grounding, and equipment installation to minimize the risk of electrical fires and accidents. Many states and local jurisdictions adopt the NEC as part of their building codes, making compliance mandatory for solar power installations.

Net Metering

Net metering is a billing arrangement between a solar power system owner and their utility provider, allowing for the exchange of excess solar-generated electricity for grid-supplied electricity. With net metering, a solar system can send surplus electricity to the grid during periods of high production, effectively crediting the owner's account. These credits can then be used to offset electricity consumption during periods of low solar production or nighttime usage, reducing overall energy costs.

Off-Grid System

An off-grid solar power system is a standalone energy system that is not connected to the main electrical grid. Off-grid systems are designed to provide all of the energy required for a home or facility, typically relying on solar panels, battery storage, and backup generators for power. These systems are common in remote locations without grid access or for users seeking complete energy independence.

Ohm

An ohm is the unit of measurement for electrical resistance in the International System of Units (SI). It represents the opposition a material offers to the flow of electric current. In solar power systems, resistance is an important factor to consider, as it can impact the efficiency and performance of electrical components, such as wires and connectors.

Open-Circuit Voltage (Voc)

Open-circuit voltage (Voc) is the voltage across a solar cell or panel when there is no electrical load connected. It is a useful parameter for characterizing solar panels, as it provides an indication of their maximum voltage output under standard test conditions. Voc is affected by temperature, with higher temperatures generally leading to lower open-circuit voltage values.

Parallel Connection

A parallel connection is an electrical wiring configuration in which multiple components, such as solar panels or batteries, are connected with their positive terminals joined together and their negative terminals joined together. This arrangement increases the overall current capacity while keeping the voltage constant, allowing for greater power output or energy storage.

Peak Sun Hours

Peak sun hours refer to the number of hours per day when the solar irradiance averages 1,000 watts per square meter, used as a standard measurement for evaluating the potential solar energy production at a given location.

Performance Ratio (PR)

The performance ratio is a dimensionless value that represents the efficiency of a solar power system by comparing its actual output to its expected output under ideal conditions. It is used to assess the system's overall performance and identify potential issues.

Photovoltaic (PV)

Photovoltaic refers to the process of converting sunlight directly into electricity using solar cells, which are semiconductor devices that generate an electric current when exposed to light.

Power Factor

Power factor is the ratio of real power (measured in watts) to apparent power (measured in volt-amperes) in an electrical circuit. It indicates the efficiency with which electrical power is being utilized and is typically expressed as a value between 0 and 1.

Power Purchase Agreement (PPA)

A power purchase agreement is a long-term contract between an electricity generator (typically a solar power system owner) and an electricity consumer or utility, where the consumer agrees to purchase the generated electricity at a predetermined price.

Power Rating

The power rating of a solar panel, inverter, or other electrical device is the maximum output power it can produce under specified standard test conditions, usually expressed in watts (W) or kilowatts (kW).

Pulse Width Modulation (PWM)

Pulse width modulation is a technique used by charge controllers to regulate the voltage and current supplied to a battery during the charging process, ensuring optimal charging efficiency and prolonging battery life.

Renewable Energy

Renewable energy refers to energy that is generated from natural resources, such as sunlight, wind, and water, which are replenished continuously and can be harnessed without depleting them or causing significant harm to the environment.

Renewable Portfolio Standard (RPS)

A renewable portfolio standard is a government-mandated policy requiring a certain percentage of electricity to be generated from renewable energy sources by utilities or electricity providers within a specified timeframe.

Series Connection

A series connection is an electrical circuit configuration in which components are connected positive to negative, end-to-end, forming a single path for the flow of current. In solar power systems, connecting solar panels in series increases the total voltage output while keeping the current constant.

Short-Circuit Current (Isc)

Short-circuit current refers to the maximum current that flows through a solar cell or module when its output terminals are directly connected, or short-circuited. This value is used to determine the electrical characteristics and performance of the solar cell or module.

Solar Access

Solar access refers to the amount of sunlight available to a specific location, taking into consideration factors such as shading from nearby buildings, trees, or other obstructions, as well as the path of the sun throughout the day and across different seasons.

Solar Cell

A solar cell is a semiconductor device that converts sunlight directly into electricity through the photovoltaic effect. Solar cells are the basic building blocks of solar modules and panels, which are used in solar power systems to generate electricity.

Solar Energy

Solar energy is the radiant energy emitted by the sun, which can be harnessed and converted into usable electricity through various technologies, such as photovoltaic (PV) systems, solar water heaters, and solar thermal power plants.

Solar Insolation

Solar insolation refers to the amount of solar radiation energy received on a given surface area during a specific period, usually expressed in kilowatt-hours per square meter per day ($kWh/m^2/day$). It is an important factor in determining the potential solar power output for a particular location.

Solar Irradiance

Solar irradiance is the amount of solar energy received per unit area, typically measured in watts per square meter (W/m^2). It varies throughout the day and across different seasons, affecting the performance and efficiency of solar power systems.

Solar Module (also Solar Panel)

A solar module, also known as a solar panel, is an assembly of interconnected solar cells mounted on a supporting structure and protected by a transparent cover. Solar modules are used to capture sunlight and convert it into electricity in solar power systems.

Solar Renewable Energy Certificates (SRECs)

SRECs are tradable certificates that represent the environmental benefits of generating one megawatt-hour (MWh) of electricity from solar energy sources. They are used in certain markets to help meet renewable energy portfolio standards and can be sold by solar power system owners to generate additional income.

Solar Resource

Solar resource refers to the amount and quality of sunlight

available at a specific location, taking into account factors such as solar insolation, solar access, and peak sun hours. It is a key determinant of the potential energy output and efficiency of a solar power system.

Solar Thermal

Solar thermal technology captures and utilizes the sun's heat energy for purposes such as heating water, space heating, or powering thermal-based electricity generation systems. Solar thermal systems typically use solar collectors, heat exchangers, and storage tanks to capture, transfer, and store the heat energy.

Solar Tracker

A solar tracker is a mechanical device that automatically adjusts the position of solar panels or mirrors to follow the sun's path throughout the day, maximizing the amount of sunlight captured and increasing the energy output of the solar power system.

Stand-Alone System

A stand-alone system, also known as an off-grid system, is a solar power system that operates independently of the electric grid, providing electricity to remote or isolated locations. These systems typically include solar panels, a charge controller, batteries for energy storage, and an inverter to convert DC power to AC power for use in appliances and devices.

Standby Losses

Standby losses refer to the energy wasted by electrical devices and appliances that continue to consume power even when they are not actively in use or are in a standby mode. Reducing standby losses can improve energy efficiency and help lower overall electricity consumption.

String Inverter

A string inverter is a type of solar inverter that converts the DC power generated by a series or "string" of solar panels into AC power, which can then be used by household appliances or fed into the electric grid. String inverters are commonly used in residential and small-scale commercial solar power systems.

System Efficiency

System efficiency refers to the overall performance of a solar power system, taking into account factors such as the efficiency of individual components, energy losses during power conversion and transmission, and environmental conditions that impact the system's ability to generate and utilize solar energy.

Temperature Coefficient

The temperature coefficient is a measure of how the power output or efficiency of a solar cell, module, or system changes with respect to temperature. It is typically expressed as a percentage change in output per degree Celsius and is used to estimate the performance of a solar power system under varying temperature conditions.

Thin-Film Solar Cell

A thin-film solar cell is a type of photovoltaic (PV) cell made from a thin layer of semiconductor material deposited on a substrate, such as glass, plastic, or metal. Thin-film cells are generally lighter and more flexible than traditional crystalline silicon cells, but they typically have lower efficiencies.

Tilt Angle

The tilt angle refers to the angle between a solar panel or solar collector and the horizontal plane. It is an important parameter for optimizing the energy output of a solar power system, as it affects the amount of sunlight captured by the panels or collectors throughout the day and across different seasons.

Transformer

A transformer is an electrical device that transfers electrical energy between two or more circuits by changing the voltage level without altering the frequency. Transformers are commonly used in solar power systems to step up or step down the voltage, depending on the requirements of the connected loads or the electric grid.

Transmission Losses

Transmission losses refer to the amount of electrical energy that is lost as heat during the transmission of electricity from power generation sources, such as solar power systems, to end users through power lines and distribution networks. Reducing transmission losses can improve the overall efficiency of an electrical system.

Voltage

Voltage, also known as electric potential difference, is the force that drives the flow of electric charge (current) through a conductor, such as a wire. It is measured in volts (V) and is a key parameter in the design and operation of solar power systems, as it affects the system's power output and efficiency.

Voltage Drop

Voltage drop is the decrease in voltage that occurs as

electrical current flows through a conductor, such as a wire or cable, due to the resistance of the conductor. In solar power systems, minimizing voltage drop is important to maintain optimal system efficiency and performance.

Watt (W)

The watt is a unit of power used to quantify the rate of energy transfer or conversion. In the context of solar power systems, the watt is commonly used to express the electrical output of solar panels, inverters, and other system components.

Watt-hour (Wh)

The watt-hour is a unit of energy equivalent to one watt of power used for one hour. It is commonly used to measure the energy production or consumption of solar power systems, electric appliances, and devices.

Wind Power

Wind power is a form of renewable energy generated by converting the kinetic energy of moving air (wind) into mechanical or electrical energy, typically using wind turbines. Wind power can be used as a complementary energy source to solar power, helping to diversify and strengthen renewable energy systems.

Zero Net Energy (ZNE)

A zero net energy building or system is one that generates as much energy as it consumes over the course of a year, typically through the use of on-site renewable energy sources, such as solar power, and energy-efficient design and technologies. ZNE buildings and systems contribute to a more sustainable built environment and can help reduce reliance on non-renewable energy sources.

[CONCLUSION] - JOIN THE SOLAR POWER REVOLUTION: BRIGHTENING THE WORLD, ONE PANEL AT A TIME

As we reach the final pages of our solar odyssey, it's hard not to feel a sense of camaraderie, dear reader. We've traversed the vast landscape of DIY solar together, delving into the nitty-gritty details and navigating the complexities of renewable energy. From the moment we embarked on this journey, you've shown unwavering curiosity and dedication to empowering yourself and greening our world, one solar panel at a time.

As you continue your journey into the world of solar energy, remember that evergreenoffgrid.com is here to support you every step of the way. If you have any lingering questions or want to dive deeper into specific topics like sizing your inverter or battery bank, our comprehensive blog articles are a great resource. Additionally, you can take advantage of our online marketplace to compare prices and shop for DIY solar components and equipment from reputable manufacturers. We're here to help make your transition to solar as seamless and rewarding as possible, so don't hesitate to explore all that evergreenoffgrid.com has to offer.

Now, like a well-prepared feast, we've laid out all the ingredients before you: the components of a solar power system, the tips and tricks for successful installation, and the financial incentives that make going solar not only environmentally friendly but economically savvy. Now, it's up to you to take these ingredients and create something truly remarkable, turning the sun's rays into a bountiful harvest of clean, renewable energy for you and your community.

As you set forth on your DIY solar adventure, remember that life, much like solar installations, can be a bit of a roller coaster. There will be ups and downs, moments of doubt, and the occasional unexpected challenge. But with each obstacle comes the opportunity to learn, grow, and become an even more ardent advocate for sustainability. Your efforts, however small they may seem, contribute to a collective movement toward a greener future for generations to come.

The pursuit of solar energy is not a solitary endeavor; it's a shared mission, driven by people like you who are passionate about making a difference. By choosing the DIY solar path, you've joined a community of trailblazers and renewable energy enthusiasts who, together, are forging a brighter, more sustainable world.

So, as we turn the final page and bid you adieu, we leave you with a heartfelt thank you for embarking on this journey with us.

We're proud to have you as a fellow traveler on the road to a greener tomorrow. Here's to all the solar-powered adventures that lie ahead, and may your pursuit of clean energy be as radiant and invigorating as the sun itself.

Farewell, intrepid solar explorer, and may the sun forever shine upon your panels.

THE END

ABOUT THE AUTHOR

Jon Springer is Seattle native with a passion for renewable energy and preserving our planet for future generations. His journey began in October 2001 when he joined the US Coast Guard, serving for five years before pursuing higher education at the University of Washington. He earned his electrical engineering degree and went on to work as a nuclear engineer for the US Navy and later as an electrical design engineer with the US Army Corps of Engineers (USACE).

His dedication to service has extended beyond his normal responsibilities, taking on collateral duties such as disaster response missions after Hurricanes including Maria and Ian, and serving as a flood team lead in his local area for USACE.

Volunteering with Kilowatts for Humanity (KWH), a group committed to ending energy poverty around the world, he contributes to the Micro-grid design team, focusing on providing community off-grid power solutions primarily in rural Africa.

As the founder of evergreenoffgrid.com, he has created a platform for DIY solar enthusiasts, offering a marketplace for comparing products from reputable manufacturers and providing educational blogs on solar system design and relevant topics.

His personal life is filled with love for his family, with whom he enjoys traveling and spending time. With such extensive experience, commitment to renewable energy, and a genuine desire to share his knowledge, Jon is thrilled for the opportunity to guide you on your journey towards energy independence.

Made in United States
Troutdale, OR
03/21/2025

29932393R00116